The Capability Approach, Technology and Design

Philosophy of Engineering and Technology

VOLUME 5

Editor-in-chief

Pieter E. Vermaas (*Delft University of Technology*): *analytic research, general topics and overarching topics*

Associated Editors

David E. Goldberg (*University of Illinois*): *practitioners' research*
Ibo van de Poel (*Delft University of Technology*): *ethics of engineering and technology*
Evan Selinger *(Rochester Institute of Technology): research in continental philosophy and science and technology studies*

For further volumes:
http://www.springer.com/series/8657

Ilse Oosterlaken • Jeroen van den Hoven
Editors

The Capability Approach, Technology and Design

 Springer

Editors
Ilse Oosterlaken
Faculty of TPM, Philosophy section
Delft University of Technology
Jaffalaan 5
2628 BX Delft
Netherlands

Jeroen van den Hoven
Faculty of TPM, Philosophy section
Delft University of Technology
Jaffalaan 5
2628 BX Delft
Netherlands

ISSN 1879-7202   ISSN 1879-7210 (electronic)
ISBN 978-94-007-3878-2   ISBN 978-94-007-3879-9 (eBook)
DOI 10.1007/978-94-007-3879-9
Springer Dordrecht Heidelberg New York London

Library of Congress Control Number: 2012934159

© Springer Science+Business Media B.V. 2012
This work is subject to copyright. All rights are reserved by the Publisher, whether the whole or part of the material is concerned, specifically the rights of translation, reprinting, reuse of illustrations, recitation, broadcasting, reproduction on microfilms or in any other physical way, and transmission or information storage and retrieval, electronic adaptation, computer software, or by similar or dissimilar methodology now known or hereafter developed. Exempted from this legal reservation are brief excerpts in connection with reviews or scholarly analysis or material supplied specifically for the purpose of being entered and executed on a computer system, for exclusive use by the purchaser of the work. Duplication of this publication or parts thereof is permitted only under the provisions of the Copyright Law of the Publisher's location, in its current version, and permission for use must always be obtained from Springer. Permissions for use may be obtained through RightsLink at the Copyright Clearance Center. Violations are liable to prosecution under the respective Copyright Law.
The use of general descriptive names, registered names, trademarks, service marks, etc. in this publication does not imply, even in the absence of a specific statement, that such names are exempt from the relevant protective laws and regulations and therefore free for general use.
While the advice and information in this book are believed to be true and accurate at the date of publication, neither the authors nor the editors nor the publisher can accept any legal responsibility for any errors or omissions that may be made. The publisher makes no warranty, express or implied, with respect to the material contained herein.

Printed on acid-free paper

Springer is part of Springer Science+Business Media (www.springer.com)

Contents

Part I Introduction

1 **The Capability Approach, Technology and Design:
   Taking Stock and Looking Ahead** 3
   Ilse Oosterlaken

2 **Human Capabilities and Technology** 27
   Jeroen van den Hoven

Part II Technology

3 **Liberation from/Liberation within: Examining One
   Laptop per Child with Amartya Sen and Bruno Latour** 39
   Kim Kullman and Nick Lee

4 **Evaluating Emerging ICTs: A Critical
   Capability Approach of Technology** 57
   Yingqin Zheng and Bernd Carsten Stahl

5 **"How I Learned to Love the Robot": Capabilities,
   Information Technologies, and Elderly Care** 77
   Mark Coeckelbergh

6 **Towards a Sustainable Synergy: End-Use Energy
   Planning, *Development as Freedom,* Inclusive
   Institutions and Democratic Technics** 87
   Manu V. Mathai

7 **Marrying the Capability Approach, Appropriate Technology
   and STS: The Case of Podcasting Devices in Zimbabwe** 113
   Ilse Oosterlaken, David J. Grimshaw, and Pim Janssen

8 From Individuality to Collectivity: The Challenges
   for Technology-Oriented Development Projects 135
   Álvaro Fernández-Baldor, Andrés Hueso, and Alejandra Boni

9 Technology Choice in Aid-Assisted Parliamentary
   Strengthening Projects in Developing Countries:
   A Capability Approach 153
   Malik Aleem Ahmed

Part III   Design

10 Design, Risk and Capabilities 173
    Colleen Murphy and Paolo Gardoni

11 Re-conceptualizing Design Through the Capability Approach 189
    Crighton Nichols and Andy Dong

12 Processes for Just Products: The Capability
    Space of Participatory Design 203
    Alexandre Apsan Frediani and Camillo Boano

13 Inappropriate Artefact, Unjust Design? Human Diversity
    as a Key Concern in the Capability Approach
    and Inclusive Design 223
    Ilse Oosterlaken

Author Biographies 245

Author Index 249

Subject Index 255

Part I
Introduction

Chapter 1
The Capability Approach, Technology and Design: Taking Stock and Looking Ahead

Ilse Oosterlaken

1.1 Introduction

The capability approach has gained popularity as a powerful conceptual framework for analyzing and discussing matters of justice, development and well-being. Technology and design are phenomena that have a tremendous influence on such matters. Scholars from different disciplines have been reflecting on these phenomena for decades, but only recently are they increasingly turning to the capability approach in order to do so. In June 2011, for example, the journal *Ethics and Information Technology* published a special issue on the capability approach and ICT.[1] This book is the first to bring together a number of papers that explore how the capability approach can be brought to bear on design and technology more broadly and – vice versa – how this may in turn feed general debates taking place within the literature on the capability approach.

In this introduction I would like to do the following. First (Sect. 1.2) I will briefly introduce the capability approach and the main concepts used in it. Readers who are new to the capability approach may find this section helpful, as it presents general theoretical background for the chapters in this edited volume. It will also provide them with some suggestions for further reading. Secondly I will – in an extensive literature review – take stock of what has been published so far on the capability approach, technology and design – what are the most interesting themes, discussions and ways of applying the approach? This will be divided in two sections: one on technology in general (Sect. 1.3) and one, more specifically, on ICT (Sect. 1.4). This is

---

[1] A concise overview of the articles in that special issue (volume 13, number 2) can be found in the editorial by Oosterlaken and van den Hoven (2011).

I. Oosterlaken (✉)
Philosophy Section, Delft University of Technology, P.O. Box 5015, 2600 GA Delft,
The Netherlands
e-mail: e.t.oosterlaken@tudelft.nl

not only meant to put the current book into perspective, but also to help new researchers in this area on their way. Thirdly (Sect. 1.5) I will introduce the chapters in this edited volume in some detail. The chapter will end (Sect. 1.6) with some reflections on the future of this new research area.

1.2 The Capability Approach

The capability approach, for which philosopher Martha Nussbaum (e.g. 2000, 2006, 2011a) and economist Amartya Sen (e.g. 1992, 1999, 2009) have done the ground work, has become an influential framework for investigating and discussing topics in the area of justice, equality, well-being and development. According to the capability approach, a key evaluative space in these areas is not income, not resources, not primary goods, not utility (i.e. happiness or the sum of pains and pleasures) or preference satisfaction. Its proponents argue that the focus should rather be on human capabilities. Capabilities are often described as what people are effectively able to do and be or the positive freedoms that people have to enjoy valuable 'beings and doings'. These beings and doings are called 'functionings' by Sen. Functionings "together constitute what makes a life valuable" (Robeyns 2005) and are "constitutive of a person's being" (Alkire 2005a). Examples of functionings are such diverse things as working, resting, being literate, being healthy, being part of a community, being able to travel, and being confident. "The distinction between achieved functionings and capabilities," so Robeyns (2005) explains, "is between the realized and the effectively possible; in other words, between achievements on the one hand, and freedoms or valuable options from which one can choose on the other." As Alkire explains, one reason to focus on capabilities instead of functionings is that we value free choice and human agency:

> A person who is fasting is in a state of undernutrition, which may seem very similar to a person who is starving. But in the one case, the fasting person could eat and chooses not to; whereas the starving person would eat if she could. (Alkire 2005b)

The capability approach thus acknowledges that people pursue not only their own well-being, but may also choose to pursue other ends; for example, the well-being of others, living up to religious ideals, or following moral norms. Hence, policies should – according to the capability approach – aim at expanding people's capabilities and not force people into certain functionings (like being well-fed).

Why should we focus on these capabilities in our developmental efforts, rather than utility or resources – including technological resources? The reason to prefer capabilities over resources as the main evaluative space is that the relationship between a certain amount of goods and what a person can do or can be varies, as Sen and others have often illustrated:

> ... a person may have more income and more nutritional intake than another person, but less freedom to live a well-nourished existence because of a higher basal metabolic rate, greater vulnerability to parasitic diseases, larger body size, or pregnancy. (Sen 1990)

Sen occasionally refers to a technical artefact to explain the same point, namely a bicycle. All bicycle owners are equal in terms of their possession of this resource, but people with certain disabilities will not gain an increased capability to move about as a result of this bicycle (Sen 1983, 1985). One could also think of other things obstructing the creation of human capabilities, such as the absence of roads – a bicycle in the Netherlands will probably add more to the owner's capabilities than a bicycle in the Sahara. One of the crucial insights of the capability approach is thus that the conversion of goods and services into functionings is influenced by personal, social, and environmental conversion factors.

The reason why capability theorists prefer these capabilities over utility or preference satisfaction is the existence of a phenomenon which Sen has called "adaptive preferences":

> Our desires and pleasure-taking abilities adjust to circumstances; especially to make life bearable in adverse situations. The utility calculus can be deeply unfair to those who are persistently deprived [...] The deprived people tend to come to terms with their deprivation because of the sheer necessity of survival; and they may, as a result, lack the courage to demand any radical change, and may even adjust their desires and expectations to what they unambitiously see as feasible. (Sen 1999)

Thus, if the deprived are happy with their lot in life we can, according to the capability approach, not conclude from this that there is no injustice in their situation. Because of these and other issues, the capability approach conceptualizes well-being in terms of a person's capabilities and development as a process of expanding these capabilities.

It should be noted here that Sen and Nussbaum use 'human capabilities' as an ethical category; the term refers to those capabilities of an individual that are ultimately or intrinsically valuable. Of course not all capabilities that a person may have belong to this category – many capabilities will only be of instrumental importance to these valuable human capabilities. Furthermore, some capabilities may be trivial from the perspective of justice and development. Sen (1987), for example, is highly skeptical about a new brand of washing powder expanding our human capabilities, as advertisers tend to claim. Moreover other capabilities may be outright undesirable to promote – Nussbaum (2000), for example, gives the example of the capability for cruelty. In short, Sen and Nussbaum agree that an ethical evaluation of capabilities is necessary – which is good to keep in mind, as quite a number of technologies may first and foremost expand undesirable, trivial or instrumental capabilities. In the latter case we should always ask to which ultimately valuable capabilities they contribute.

Not surprisingly, then, one important debate within the capability approach is about which capabilities matter and who (how, when) is to decide this. This is actually one of the main topics on which Sen and Nussbaum differ of opinion.[2]

---

[2] For an analysis of the differences between the capability approach of Nussbaum and Sen, see for example Robeyns (2005) and part II of Crocker (2008).

Nussbaum has, after extensive discussion with people worldwide, composed a list of 10 central human capabilities that are needed for living a worthwhile life in conformity with human dignity. She claims that justice requires bringing each and every human being over a certain threshold for each of the capabilities on her list. Although Sen gives plenty of examples of important capabilities in his work, he has always refused to make such a list. His reasons are that the proper list of capabilities may depend on purpose and context, and should be a result of public reasoning and democracy; not something a theorist should come up with. Yet it is being recognized by both Sen and Nussbaum that from an ethical perspective not only outcomes in terms of expanded capabilities matter, but also the process through which these changes are brought about (e.g. its fairness). Democracy, public deliberation and participation are – because of the 'list debate' and because of the value attached to human agency – also frequent topics of reflection and discussion amongst other capability theorists (see e.g. Crocker 2008).

Various other topics and questions also feature a lot in the literature on the capability approach and in this short introductory section we can mention only but a few of them. One is, not surprisingly, the question of how to operationalize the capability approach (see e.g. Comim et al. 2008). As Alkire (2005a) explains, "operationalizing is not a one-time thing," but something that is dependent upon such things as country, level of action and the problem at hand. This raises important questions on how to identify, rank, weigh, trade-off relevant capabilities in policy/project applications, on which no consensus exist. One of the many challenges is also that it is hard to measure capabilities, as they (a) refer to the possible and not just to the realized and (b) are a complex construct depending on both an individual's internal characteristics/capacities and his/her external environment. A challenge is furthermore how to 'aggregate' over people while not loosing sight of the fact that a capability approach emphasizes that each and every person needs sufficient capabilities to lead a flourishing life. Another topic of discussion has been whether or not the capability approach, with its emphasis on an individual's capabilities, is not too individualistic and pays enough attention to groups and social structures (see e.g. Robeyns 2005). Some authors have argued that groups should be given a more central role (e.g. Stewart 2005) or that the framework of the capability approach should be extended to include collective capabilities (e.g. Ibrahim 2006).[3]

Not only has the capability approach had an influence on scholarly work, but it has also influenced policy and practice. It is, for example, well known that it has provided the intellectual foundations for the human development paradigm of the United Nations Development Program (UNDP), which is reflected in their annual Human Development Reports.[4] The capability approach has been applied in different ways (Robeyns 2006), such as the assessment of small scale development projects

---

[3] This debate is also addressed at several points in a recent edited volume by some leading scholars on the capability approach (Comim et al 2008).

[4] For an introductory textbook on the capability approach and human development, see the edited volume by Deneulin and Shahani (2009).

1   The Capability Approach, Technology and Design: Taking Stock and Looking Ahead   7

(this could also include projects involving technology), theoretical and empirical analyses of policies (this could also involve technology policy or perhaps new technologies) and critiques on social norms, practices and discourses (one could think here of design practices and the 'ICT for Development' discourse). Many of the applications so far have been concerned with assessment and evaluation, but of course for advancing justice and development 'prospective' applications should also receive attention (Alkire 2008), meaning that we should investigate how the expansion of human capabilities can successfully be brought about. In general terms:

> For some of these capabilities the main input will be financial resources and economic production; but for others, it can also be political practices and institutions, […] political participation, social or cultural practices, social structures, social institutions, public goods, social norms, traditions and habits. (Robeyns 2005)

Technology and design could, of course, also be important inputs for the expansion of valuable capabilities and it is thus to be applauded that increasing attention is paid to this topic. For example, in September 2009 the thematic group 'technology and design' was established under the umbrella of the Human Development and Capability Association (HDCA). On the association's website (www.hd-ca.org) you can find more information on this group and its members. The website also presents a general introductory reading list and a more extensive bibliography on the multi-disciplinary body of literature on the capability approach, with publications ranging from philosophical essays to empirical analyses. The reader is referred to this reading list/bibliography for more information. I will now discuss what has appeared so far on the more specific topic of the capability approach and technology.

1.3   Applying the Approach to Technology – Taking Stock

To say that technology is – in various ways – relevant for human development, justice, equality, well-being and empowerment is to force an open door. Yet the capability approach, despite being an influential framework for critical engagement with such issues, has not been applied that often to technology – or at least not in an explicit and extensive way.[5] One of the first authors to do so was – as far as I know – Garnham (1997, 2000). He applied the capability approach to the topic of ICT policy. "Thinking of entitlements in terms of functionings and capabilities", he argued convincingly, "allows us to get behind the superficial indices of access and

---

[5] This is not to say that nobody has ever touched upon the topic, on the contrary. Sen and Nussbaum have both mentioned the potential of technology for expanding human capabilities. A recent textbook, *An Introduction to the Human Development and Capability Approach* (Deneulin and Shahani 2009), says Birdsall (2011), "provides numerous examples of the impact of ICT on various social and economic public policy sectors crucial to the enhancement of human capabilities." Many of the annually published *Human Development Reports*, for which the work of Sen has provided the intellectual foundation, have also paid attention to technology. See also a recent UNDP report written by Hamel (2010). And Qureshi (2010), for example, brings up the capability approach in a recent editorial for the journal *Information Technology for Development*.

usage that we so often use." For some years to follow only a handful of writings seems to have followed suit. Since roughly 2006, however, there seems to be a steep increase in the number of scholars using the capability approach for deliberations about technology. Just as is the case with the literature on the capability approach in general, these scholars come from different disciplines, like development studies, economics, science and technology studies, philosophy of technology and computer science. Their publications are scattered over many journals and books and sometimes do not refer to related publications in this area, as if some authors are inventing the same wheel independently from each other. Hence, it seemed useful to include an extensive literature review[6] in this introduction, which will not only put the current book into perspective, but hopefully also help new researchers to find their way in this upcoming area.

One thing that is salient when going over the literature that has appeared so far in this area, is that a majority of publications is concerned with ICT – I will get back to this in section three. Yet several authors have addressed other technological domains, specific engineering issues or *technology and the capability approach in general*. To start with the latter, Oosterlaken (2011) wrote a reply to an article (Smith and Seward 2009) that extensively discusses individuals and social structures as the constituents of human capabilities. She explains that in such an ontology technical artefacts should be explicitly acknowledged as an important third component. Referring to Actor-Network Theory (e.g. Latour 2005), she furthermore argues that technology only expands human capabilities when appropriately embedded in wider social and physical structures. Sen himself (2010) recently brought up for discussion the idea that different artefacts may have – by their nature – different impacts on human capabilities. He points out that all of them can, of course, in principle be used for good or for bad – like a gun for protecting innocent lives and a mobile phone for planning a terrorist attack. Yet:

> A telephone owned by a person helps others to call the person up, as well as to receive calls from him or her, and so the increased freedom of the phone owner adds to the freedom of others. In contrast, a gun owned by one can easily reduce the freedom of others, if the gun is pointed at them—or could be. Many goods have little impact on others, as a shirt owned by one does not, typically, have much of an impact on the lives of others. (Sen 2010)

Thus, Sen concludes in his article titled *The Mobile and the World*, the phone "is generally freedom-enhancing, and that is an appropriate enough point of departure for the hagiography of the mobile phone." Whether or not there is indeed such a principled difference between phones and guns and how this can be accounted for might be an interesting topic of further reflection by philosophers of technology. Coeckelbergh (2011) has criticized the view in which technologies are seen as mere

---

[6] Considering the rate at which new publications recently have appeared – and continue to appear – at many different places, I do not claim completeness. Without doubt there are publications which have escaped our attention. For example, articles in other languages than English have not been scrutinized on relevance. It should furthermore be noted that conference papers have not been included in this overview, even though some (like Gigler 2004) are interesting and relevant. Moreover, we have been selective in leaving out publications that only loosely mention the capability approach without discussing the approach or any of its main ideas/concepts in any detail.

instruments for the expansion of timeless and universally valid central human capabilities. He makes a plea for

> a hermeneutics of techno-human change, involving interpretations of dynamic relations between unstable capabilities, technologies, practices, and values. This requires us to use the capability approach in a way that highlights its interpretative dimension.

This ties in with the debate about the universality and validity of Nussbaum's list of ten central human capabilities, which Coeckelbergh addresses towards the end of his article. A general discussion of the place of technology in the capability approach and some problems and prospects can furthermore be found in a book chapter by Johnstone (2012). She draws – amongst others – attention to the fact that the relation between technology and human capabilities is not as simple and straightforward as one might think. For example "one and the same technology may have both capability enhancing and capability diminishing effects even for the same person" and "technology is also deeply entangled in the broader social and material context", which may indirectly affect well-being and agency.

Central in the development of new technological artefacts and to the job of many engineers is *design*. A capability approach of design has been most extensively advocated by Dong (2008) and Oosterlaken (2009a). These two articles each take a different angle and nicely complement each other. Dong applies the capability approach to design policy and the process of design. He argues that from a justice perspective we should pay attention to citizens' capabilities to design and in this way co-shape their life world. Oosterlaken emphasizes that design outcomes or the design features of an artefact are very important for its exact impact on human capabilities. Hence, she introduces the idea of 'capability sensitive design', analogue to the idea of 'value sensitive design' originating from the field of ethics of technology (e.g. van den Hoven 2007). Toboso (2011) argues that designers should pay more attention to human diversity, as we saw a core theme in the general literature on the capability approach. He discusses the link between the universal/inclusive design movement and the capability approach and introduces the concept of "functional diversity" to support the shift in design practice that he proposes. Architects and urban designers may be interested in the work of Frediani in the area of housing and settlement upgrading projects in developing countries. "Urban programmes with the objective of enhancing people's freedoms should", he says (Frediani 2007), "be engaged in identifying the underlying physical designs of the built environment." Based on a case study, he argues that the World Bank fails – in several ways – to genuinely expand valuable capabilities of squat dwellers, although the word 'freedom' figures strongly in its policy language.[7]

In all engineering domains *risk* is a recurring issue that often raises difficult ethical questions, for example with respect to informed consent and a just distribution of such risks (e.g. Asveld and Roeser 2009). Gardoni and Murphy – a civil

---

[7] A practical application of the capability approach to this field is also mentioned by Rubbo (2010). She says that the approach, with its ideas about agency, has been a source of inspiration for the international design program Global Studio. Unfortunately, the article lacks detail on how exactly the capability approach has made a difference in this program.

engineer and a philosopher respectively – have together written a series of articles in which they develop, step-by-step, a capability approach of risk assessment and management. They discuss the shortcomings of existing risk approaches, including cost-benefit analysis (Murphy and Gardoni 2007) and argue that risks should be assessed according to a hazard's impact on individual capabilities (Murphy and Gardoni 2006). Topics addressed by them include minimum thresholds for capabilities and the acceptability or tolerability of risks (Murphy and Gardoni 2008), a Disaster Impact/Recovery Index analogue to the UNDP's Human Development Index (Gardoni and Murphy 2008), and a plea for a focus on capabilities instead of functionings in risk analysis (Murphy and Gardoni 2010).

The capability approach has also been used to reflect on issues raised by new *bio-technologies and health care technologies*. Cooke (2003) argues that the capability approach "can be used as a framework to ensure freedom and equality in the use of germ-line engineering technology." In her article she compares Sen's capability theory with Norman Daniels "normal functionings model." Clague (2006) discusses the capability approach in relation to the commercialisation of bio-technologies and "patent injustice." With respect to bioethics, Nussbaum (2008) has argued amongst others that it is important that capabilities and not actual functioning is the political goal of bio-technologies. She furthermore notes that referring to the "natural" in such debates is unhelpful, as sports like skiing also depend on the usage of all sorts of artefacts for an "unnatural" enhancement of our capabilities. It ties in with before mentioned article of Coeckelbergh (2011), who likewise believes that human enhancement[8] is not fundamentally different from capability expansion with more traditional technologies. Finally, some authors have used the capability approach to reflect on the usage of robots in health care. Coeckelbergh (2009) argues that Nussbaum's capability list provides specific, positive criteria to evaluate this technology in relation to the quality of care. Referring to his work, Borenstein and Pearson (2010) address the issue of human needs and human-robot interaction in different health care contexts and the demands that this makes on the careful design of these artefacts.

Although a bit outside the scope of this book, it may be worth saying something about innovation studies and the capability approach. Cozzens et al. (2007) sketch a broad overview of how (a) science and technology studies, (b) economic growth theory and (c) innovation systems research have approached the topic of development. None of these, they say, "explicitly takes development-as-freedom [i.e. development conform Sen's capability approach] as its goal nor explores concretely how the approach would contribute to meeting the basic needs of the world's population." Scholars in these fields should however do so, the authors argue, in order to explicate

---

[8] Coeckelbergh describes human enhancement as follows: "Human enhancement aims at using technology to create better humans. What this means can best be clarified by saying what it is not: its aim is not therapeutic: it does not restore humans to a 'normal' state but wants to create humans that are 'better than normal', 'better than human'." Human enhancement is closely associated with new and emerging technologies like neurotechnology and nanotechnology.

the contribution of their respective field to the development agenda. Musa (2006),[9] however, does connect the capability approach to work from innovation studies more broadly. He adjusts the well-known 'Technology Acceptance Model', in which the main variables are perceived usefulness and ease of use, in order to better suit the situation in developing countries. The revised model, he claims:

> recognizes the fact that technology acceptance or adoption is ultimately influenced by the values that individuals place on technology in their daily lives and that these accumulated values will allow a country to realize the full impact of technology for development. (Musa 2006)

I will now turn to a review of the literature on ICT and the capability approach. However, some of the points made in the next section also apply to technology more broadly, just like some of the subjects discussed in this section – for example, on capabilities and design – are also relevant for ICT.

1.4 The Capability Approach and ICT

The articles that have appeared on ICT and the capability approach are many by now and can unfortunately not all be discussed separately in this introductory chapter. Many of them are focused on developing countries, but not all of them (e.g. Garnham 2000; Mansell 2001). They vary from theoretical or philosophical reflections – to which the present book is also inclined – to applied case studies. Several of the articles make a strong general case for applying the capability approach to ICT (e.g. Garnham 2000; Mansell 2001; Alampay 2006a, b; James 2006; Thomas and Parayil 2008). But as a first introduction to the topic the articles of Johnstone (2007) on the capability approach and computer ethics and of Zheng (2009) on the capability approach and ICT4D are especially recommendable. Both these articles extensively discuss the advantages and challenges of applying the capability approach to ICT and present an overview of research questions that deserve further attention. Several other articles explicitly address – with a more empirical focus – the question how the capability approach can be operationalized in relation to ICT. Only Barja and Gigler (2007), Alampay (2006b) and Wresch (2007, 2009) do this at a meso or macro level, the other authors aim to develop a framework for evaluating ICT4D projects at the micro level (e.g. Gigler 2004, 2008; Kleine 2010; Grunfeld et al. 2011; Vaughan 2011). Many articles may furthermore be read with a specific interest in the sort of ICT or application that they discuss, like mobile phones (e.g. Sen 2010), the internet/websites (e.g. Wresch 2007, 2009), telecentres in deprived regions (e.g. Garai and Shadrach 2006; James 2006; Ratan and Bailur

---

[9] Von Tunzelmann and Wang (2007), in an article on the theory of production, also use the work of Sen "to match the heterogeneity of products and their characteristics existing in markets to the heterogeneity of consumers and their demands." Yet they stay within their own discipline and do not put it in the perspective of larger issues of poverty and development, as Cozzens et al seem to have in mind.

2007; Thomas and Parayil 2008; Grunfeld et al. 2011; Vaughan 2011), e-governance systems (e.g. Madon 2004; Ahmed 2011), before mentioned care robots (e.g. Coeckelbergh 2009; Borenstein and Pearson 2010) and ICT systems in the health care sector (e.g. Zheng and Walsham 2008).

As mentioned, ICT is dominant in the literature that has appeared so far on the capability approach and technology. And many of these publication more specifically focus on 'ICT for Development' (ICT4D). One can only speculate why this is so, but at least two factors may play a role. Firstly, ICT has become extremely popular as a 'weapon against poverty' in the last 10–15 years. The example of farmers in developing countries being able to acquire crop prices by means of a mobile phone and hence being able to raise their income is by now quite well-known. The enthusiasm for ICT4D has given rise to a critical 'countermovement' and some of these authors have used the capability approach for voicing their criticism. For example, it has been claimed that too much emphasis has been put on mere resource distribution or ICT access (e.g. Madon 2004; Alampay 2006a, b; Hellsten 2007; Zheng 2007). From the perspective of the capability approach this often does not lead to positive development outcomes for everybody, as a great variety in conversion factors (see Sect. 1.2 for an explanation of this term) exist. It could even be the case (Thomas and Parayil 2008) that ICTs increase inequality, as the socially advantaged classes in a society may be more able to convert access to ICT into something useful in their lives than the already deprives ones. Kleine (2010) claims that the mainstream discourse on ICT4D "remains heavily focused on economic growth, which is too narrow to capture the impacts of ICT." Zheng (2009) furthermore notes that ICT4D often treats people as "passive receivers" of new technologies that are supposedly good for them, while the capability approach values agency and would hence take their felt needs and aspirations into account. All these authors find a powerful conceptual framework in the capability approach, which can be used to fruitfully reflect on ICT4D. Their work fits in with some of the sentiments in the wider ICT4D community. For example, one of the conclusions of Walsham and Sahay (2006) in their overview of 'information systems research in developing countries' is that there is a need for "more emphasis in future work on the *meaning* of development, and how ICTs link to this" (emphasis is mine). Heeks – a prominent scholar in the area of ICT4D – recently (2010) stated that

> the main practical call […] is still for more theory-based evidence about ICTs' impact on development; especially for more evidence founded in theories that have currency within development studies.

The capability approach is such a theoretical view on the meaning of development that has become quite influential in development studies.

A second factor in explaining the dominance of ICT in the 'technology and capability approach literature' could be that ICT seems to have – contrary to many other technologies – a quite indeterminate character, in the sense that it can directly and simultaneously contribute to the expansion of human capabilities in very different areas: health, education, recreation, livelihoods, democracy, etc. ICTs might thus be seen as the ultimate embodiment of the ideal of the capability approach that we ought to promote a variety of capabilities and leave it up to empowered individuals which functionings to realize, depending on their idea about the good life. Whether ICTs can and do fully live up to that promise can, however, be challenged from both practical experiences and philosophical insights. This point has been made most explicitly by Kleine (2011). "Due to the multi-purpose, multi-choice nature of the internet, this area of development studies", she says, "is particularly well-suited to be a test-case for the choice paradigm in development evaluation, execution and planning." Unfortunately:

> the structure of the 'development industry' is such that funders tend to be persuaded to commit resources based on the promise of pre-determined development outcomes, not by a promise that people will be empowered to make much less predictable choices of development outcomes. (Kleine 2011)

An interesting case with respect to this issue, namely so-called rural telecentres where villagers can get access to all sorts of ICT, is analyzed by Ratan and Bailur (2007). The implementing development organisations intend these centres to contribute to pre-defined well-being goals in areas like education, livelihoods or health care, whereas villagers – just like people in the West – often tend to use the telecenters for entertainment, desktop publishing and so on. This may obviously raise a dilemma for the development organisation in question. The capability approach, so these authors note, is able to theorize this. One of the problems may be an overestimation on the part of development organizations of what villagers can do and be with ICTs and the information to which they give access:

> We do not claim that people are not interested in their own welfare, but that this value is hard to see and turn into tangible welfare gains in ICTD projects, given the numerous factors that influence the translation of welfare information into welfare outcomes in developing country contexts today. (Ratan and Bailur 2007)

Furthermore, Ratan and Bailur argue that what a rational usage of ICTs is, may be quite different from the perspective of people living in great poverty and uncertainty.

Just as for technology in general, not only the use but also the design of ICT should be a subject of scrutiny from the perspective of the capability approach. Here as well agency and well-being are values that should be considered, but could contradict. Kleine (2011), for example, introduces the concept of a 'determinism continuum' on which ICTs could be placed "based on the degree to which the spectrum of user choices is already pre-determined by the technology." She thus recognizes that the design features of ICTs matter and that not all technologies are equally good from the perspective of the capability approach; some of them may restrict agency more than others. The pre-determination of choices could, for example, be based on ideas of designers about what fosters well-being. The relevance of design features also implies that deliberation about technology choice is important. A recent article by Zheng and Stahl (2011) also adds a critical note to the idea that ICTs are per definition a powerful tool to expand people's agency. Critical theory, they claim, helps to reveal and address that technology is implied in the distribution of power and sometimes in oppression and therefore possesses "ideological qualities." If one wants to expand human capabilities and agency with the help of ICT, these authors make clear, one should look into the design and regulation of technology. For that purpose the capability approach could learn from 'critical theory', which shares the value that the capability approach attaches to empowerment and which has a rich history of engaging with technology and ICT. Yet Zheng and Stahl feel that critical theorists sometimes get stuck in their attempt to "debunk positive myths" about technology. The capability approach, however,

> by seeing ICT as means to development and asking questions about what conversion factors need to be in place to facilitate the achievement of potential freedom that technology provides.

provides a counterpoise to that tendency.

Some other articles on ICT and the capability approach are also worth mentioning separately in the context of this book. The article of Johnstone (2007) makes an argument that ICTs are increasingly part of complex systems, while ethical theory is traditionally mostly action/agent oriented. This system character of ICTs makes ethical analyses in terms of solely *individual* actions, agents, intentions, reasons and obligations difficult. Thus, she concludes, a value-based approach is necessary. The capability approach, with the central importance it gives to agency and valuable human capabilities, can be used for this purpose. It can, Johnstone claims, incorporate the system level effects of ICTs through their influence on the social and material environments that influence the conversion of resources into capabilities. For Foster and Handy (2008) ICTs make the fundamental question on what different types of capabilities exist more salient. They argue that ICTs can "dramatically amplify" people's 'external capabilities', the capabilities that one has because one is directly connected to other people with certain capabilities. An example would be a farmer that can increase his income because his neighbour has an internet connection and hence access to relevant agricultural information. Van den Hoven and Rooksby (2008) have addressed the value of information in relation to distributive justice, a topic that has obviously gained in urgency as a result of the increasing prominence of ICTs in our societies. They argue that access to information is a Rawlsian primary good, but also believe that "any attempt to extend Rawls' theory of justice to information goods should take Sen's concerns into account." After all, they ask, "what value do information liberties have for someone with a substantial mental impairment, or for someone who is physically incapabile of using a computer?" Finally, Birdsall (2011) and Hellsten (2007) share their concern about how features of our modern world, such as globalisation, mass media and free markets, work out for the contribution of ICTs to human capabilities. Birdsall (2011) focusses on the connection between capabilities and human rights (see e.g. Sen 2004, 2005). He meticulously explores the differences and parallels between the literatures on the capability approach and on the 'right to communicate' (RTC). Freedom of speech and other classical negative communication rights, he argues, in our current world insufficiently guarantee people's capabilities for true communication and participation in societal dialogues. Hellsten (2007) acknowledges – like many others – the potential of ICTs for the expansion of human capabilities, but also notices that ICT needs to be used wisely if we are to achieve social justice. "The more efficient and equal use of ICT", she says, "cannot be detached from the civic, professional and ethics education."

1.5 About the Chapters in This Book

To the fastly growing body of literature discussed in the previous two sections, this volume adds 12 new contributions. As one might expect from an edited volume within a book series on philosophy of technology, the chapters are all qualitative in nature, focusing on conceptual issues and/or case study analysis.[10] Furthermore, just as in the literature that has appeared so far, ICT is also dominant in this edited volume. Two chapters discuss quite recent and high-tech forms of ICT relevant mainly in a Western context, namely care robots (Coeckelberg, Chap. 5) and ambient intelligence, affective computing and neuroelectronics (Zheng and Stahl, Chap. 4). Two other chapters assess more conventional forms of ICT applied in a development context, namely podcasting devices and computers as part of telecentres (Oosterlaken et al., Chap. 7) and television and the internet as part of e-governance initiatives (Ahmed, Chap. 9). An ICT artefact designed especially for usage in the South, namely as part of the One Laptop Per Child project, features in the contribution of Kullman and Lee (Chap. 3). However, other engineering disciplines also play a role in this volume. Energy technology is being discussed in two chapters, in one more at a policy level (Mathai, Chap. 6) and in another at the level of development projects (Fernández-Baldor et al., Chap. 8). And in the part on design there is a chapter that has direct relevance for civil engineering (Murphy and Gardoni, Chap. 10) and one that extensively addresses architecture/urban design (Frediani and Boano, Chap. 12). As the chapters are all preceded by an abstract, I will not summarize their contents here one by one. Rather, I would just like to pick out central cross-cutting concerns and highlight recurring themes in the different chapters.

The nature of the relation between technology and human capabilities in general is being addressed in a number of chapters in this volume, such as in the introductory contribution by van den Hoven (Chap. 2). Technical artefacts could, he explains, be conceptualized as "agentive amplifiers", as "they create possibilities we would not have without them." From a more anthropological perspective – drawing on the work of the Spanish philosopher of technology Ortega y Gasset – he furthermore argues that humans have always created technologies with the aim of "contributing to people's capabilities to lead flourishing human lives." Oosterlaken (Chap. 13) explains – adopting recent work in analytic philosophy of technology – that human what allegedly distinguishes technical artefacts from other types of artefacts is their being embedded in a use plan, which specifies the conditions that need to be fulfilled and the actions that need to be undertaken by the user to reach a certain goal. Enabling users to reach that goal – by expanding their capabilities – is what engineering design is ultimately all about (and not about creating material gadgets per se). A more continental approach to philosophy of technology is taken by Coeckelbergh (Chap. 5). By sketching a scenario about future robotic elderly care and the capability of social affiliation, he makes clear how "particular technologies change

---

[10] Yet more macro/quantitative issues with respect to the CA and technology could certainly be useful for informing policy and is quite conceivable, considering that the literature on the capability approach at large addresses this type of work extensively.

the meaning of the capabilities" on Nusbaum's list. His example illustrates how moral imagination, which is also important in the work of Nussbaum, can be very helpful for the hermeneutics of techno-human change that he proposed in earlier work (Coeckelbergh 2011). In his view "this ethical-hermeneutical interpretation and use of the capability approach" may be less "solid", but it results in a "more adequate conceptual instrument" for exploring the influence of technology on our lives.

One thing that is noteworthy is that many of the contributors to this edited volume discuss the capability approach in combination with existing prominent theories and perspectives with respect to technology and design, such as universal or inclusive design, appropriate technology, critical theory of technology, actor-network theory and participatory design. This is to be expected and encouraged, since – as Robeyns has noted – the capability approach

> is not a theory that can explain poverty, inequality or well-being [...] Applying the capability approach to issues of policy and social change will therefore often require the addition of explanatory theories. (Robeyns 2005)

What the capability approach can do (Robeyns 2005) is provide "a tool and a framework within which to conceptualize and evaluate these phenomena. However, "the capability approach offers little about understanding details of technology and their relationship with social processes" (Zheng 2007), or about understanding details of how technology and human capabilities are related. One example of combining the capability approach with other perspectives is the contribution by Oosterlaken (Chap. 13), who identifies the saliency of human diversity as a shared core insight of both capability theorists and the *universal/inclusive design movement* (as Toboso 2011 also did). It is when designers take the personal conversion factors of different groups of users – like disabled people or senior citizens – properly into account, that technical artefacts contribute to the expansion of the capabilities of as many people as possible. A well-known example of inclusive design are public buildings that are accessible by both stairs and a wheelchair ramp. When it comes to the practical realization of the moral ideals of the capability approach, Oosterlaken says, one could learn a lot from this design movement. At the same time, an ethical account like the capability approach is needed to deliberate about the question if a case of exclusive design is a case of injustice.

The so-called *appropriate technology movement*, to give another example of a relevant perspective on technology, is being discussed in the chapters of Fernández-Baldor et al. (Chap. 8) and of Oosterlaken et al. (Chap. 7).[11] The latter emphasize the commonalities with the capability approach. As they see it, the appropriate technology movement has always been aware of the importance of human diversity and social and environmental conversion factors influencing the impact of technical resources on development, even though it uses a different language from the capability approach. Both their case and that of Ahmed (Chap. 9) illustrates that the prevailing conversion factors may be such that a choice for a

---

[11] Nichos and Dong (Chap. 10) also mention the appropriate technology movement, I will get back to that.

1 The Capability Approach, Technology and Design: Taking Stock and Looking Ahead   17

relatively 'low-tech' solution may be preferable in ICT4D projects.[12] Fernández-Baldor et al. argue, however, that adopting a capability approach would go beyond taking the approach of the appropriate technology movement. Amongst others it would, the authors say, mean paying more attention to the process of change – which should be participatory – and expanding individual and collective agency along the way. They give an example of a local power plant project in Latin America that was not only successful in terms of sustainable energy supply, but also led to an increase in collective agency of the community in question. Their findings have some resonance with those of Gigler (2008) concerning ICT4D projects. According to him

> frequently, the most immediate and direct effect of ICT program seems to be the psychological empowerment of poor people, whereby newly acquired CT skills provide the marginalized with a sense of achievement and pride, thus increasing their self-esteem.

Thus what matters is not only what the technology directly enables people to do, but also what the introduction of the technology brings about in a wider sense. One interesting question for future research then becomes when such positive 'side effects' occur, under what circumstances technology-oriented development project contribute optimally to increasing self-esteem or individual/collective agency.

Critical theory – having influenced thinking on urban planning and architecture – is used by Frediani and Boano (Chap. 12) to reflect on power and empowerment in the production of spaces. How critical theory – which is also part of both science and technology studies (STS) and information systems research – can add value to the capability approach is also taken up by Zheng and Stahl (Chap. 4). They believe that "the capability approach can be enriched" by critical theory in three ways: "the conception of technology, the conception of agency and its 'situatedness' and the methodological implications of these two conceptions." Their reflections lead them to formulate four principles as part of a 'critical capability approach of technology', namely the principles of:

1. human-centred technological development,
2. human diversity,
3. protecting human agency and
4. democratic discourse.

Zheng and Stahl illustrate these principles by applying them to some emerging ICTs. *Science and technology studies (STS)* is, furthermore, a source of relevant insights for Kullman and Lee (Chap. 3) and – again – Oosterlaken et al. (Chap. 7). From STS these chapters highlight Actor-Network Theory and the work of Bruno Latour for understanding how human capabilities and human agency come about in and are shaped by a network of interdependencies between both humans and technical artefacts. Kullman and Lee discuss this most thoroughly, meticulously comparing the work of Sen and Latour and applying the insights gained from this to the case of

---

[12] Note though that Ahmed does not refer to the appropriate technology movement.

One Laptop per Child.[13] Oosterlaken et al. address STS insights in a much more sketchy way, it being but one part of the chapter in which an ICT4D project is being discussed from different angles.

The chapters mentioned so far already show that we can find central concepts and themes within the capability approach – most prominently human diversity, agency and participation – throughout this book as well. Participation, for example, is implied in Zheng and Stahl's 'principle of democratic discourse' about technology and emphasized by Fernández-Baldor et al. in their critical discussion of technology transfer projects. Participation – more in particular, participation in design – is also centrally addressed in the contributions by Nichols and Dong (Chap. 11) and Frediani and Boano (Chap. 12). The former follows up on an article by Dong (2008). Instead of bringing people appropriate technology, Nichols and Dong believe, we should enable them to create their own artefacts and material surroundings. In line with the capability approach, gaining design capacity or skill – as targeted by the 'humanitarian design community' – is, in the view of these authors, not enough for truly gaining the 'capability to design'. The latter may, for example, be inhibited by political factors even though design skills are present. They also acknowledge the importance of culture and briefly present a re-interpretation of Dong's 'design capability set' from the perspective of the First Australians. In Chap. 12 Frediani and Boano, who focus on urban design, note "a suprising lack of literature investigating the conceptual underpinnings of participatory design and its implications in terms of practice", a gap which the capability approach may be able to fill. They aim to overcome the "unhelpful dichotomy" between a product-oriented (e.g. Oosterlaken 2009a) and a process-oriented (e.g. Dong 2008) capability approach of design:

> …the analysis should not merely engage with the process of design, but also with its outcomes. The reason is that citizens' design freedom is shaped not merely by their choices, abilities and opportunities to engage in the process of design, but also by the degree to which the outcomes being produced are supportive of human flourishing. (Frediani and Boano)

Both the traps of physical determinism and social determinism should, according to these authors be avoided – and they criticize Dong (2008) for falling into the latter by "paying little attention to the physical and spacial outputs generated by the process of design."

Also addressing design – from a completely different angle – is Chap. 10, in which Murphy and Gardoni continue their previous work on the capability approach and technological risks, with a focus on civil engineering. They note that the existing

> reliability-based design codes only focus on probabilities and ignore the associated consequences […] there is a need for a risk-based design that accounts in a normative and comprehensive way for the consequences associated to risks.

---

[13] The case of One Laptop per Child in Australia is also briefly addressed by Nichols and Dong in Chap. 11.

The capability approach is, according to them, able to fullfil this need. Advantages include that

> a capability-based design can provide some guidance to engineers as they make tradeoffs between risk and meeting other design constraints, some of which may be also translated in terms of capabilities.

They also find it a "central principled advantage" that the capability approach "puts the well-being of individuals as a central focus of the design process" – something which may be easily lost out of sight in the practice of creating infrastructural works. Like Zheng and Stahl, they thus find it an advantage that the capability approach would promote a human-centered view on technological development.

Finally, Mathai (Chap. 6) defends that choices in energy technology and infrastructure should be more responsive to deliberation about ends and values. He sees a role here for the capability approach, as it provides a rich information basis and vocabulary to discuss the "end-uses" of energy, namely the functionings and capabilities to which energy could contribute. In light of sustainability it is inevitable that this is turned into a normative exercise, so Mathai believes, in which valuable capabilities and functionings are distinguished from the trivial and negligible ones. Yet as sustainability problems are the collective outcome of individual choices, the debate between capability theorists on individual agency and the structures of living together comes to the fore. Although the chapter is less directly engaged with technology and design itself than many of the other contributions, it is a nice example of how the capability approach can be used to address the broader implications of technological progress.

1.6 Looking Ahead

Technology and design play an important role in matters of justice, development, and our quality of life. The capability approach offers a powerful and influential conceptual framework to analyze and discuss these matters. With this book an attempt is made to further stimulate research on the capability approach, technology and design and to strengthen the emerging community of scholars working in this area. Leading theorists on the capability approach – like Robeyns (2005) and Alkire (2005a) – rightly consider its application and operationalization to be a highly interdisciplinary exercise and thus this research community will need to consist of people from many different fields. It should include scholars from philosophy and ethics of technology, science and technology studies, information systems research, engineering design, as well as social scientists working on the various domains of human life and society in which technology plays a role (such as health, governance, and so on). A key aim of this research community should be – to borrow the words of Alkire (2005a) – to 'trace the implications of the capability approach all the way through' for the area of technology and design, both in theory and practice. The rapid developments during the past couple of years on this topic indicate that the timing is right for this sort of research to take off ground. Several converging academic trends may be seen to contribute to this. For example, an 'empirical turn' has occurred since the 1980s within philosophy and ethics of technology, meaning that it is now commonly held that "philosophical reflection should be based on empirically adequate descriptions reflecting the richness and complexity of modern technology" (Kroes and Meijers 2000). Moreover there has been, so my co-editor van den Hoven explains in the second introductory chapter, an 'ethical turn' in engineering design and a 'design turn' in applied ethics. The stage thus seems perfectly set for technology, design and the capability approach to also connect more intimately.

So what sort of research questions should be answered and what sort of studies should be undertaken? I will not attempt to develop a systematic or extensive research agenda here. This is something that should still crystallize in the upcoming years, in interaction between the researchers mentioned in this introduction and others. One general remark that one could make about this, is that it would be good to pay attention to the insights and lessons that can be derived from different domains and sectors to which the capability approach has already been applied and in which technology and design also play a role (such as health, education, etc). Furthermore, much more empirical work should take place on all levels, ranging from micro to macro, in response to challenges faced by practitioners, policy makers and activists. This book hopes to contribute to some theoretical and conceptual foundations for doing so.

Johnstone (2007) and Zheng (2009) have already provided a good basis for further developing a research agenda. The general questions that Zheng has formulated with respect to ICT4D and the capability approach are categorized in four groups: (a) means and ends of development, (b) commodities, capabilities and human diversity, (c) agency and situated agency, and (d) evaluative spaces. They can be found in Table 1.1. Johnstone discusses a capability approach research agenda for computer ethics more broadly. Some examples of direct and indirect capability-related questions formulated by her can be found in Table 1.2. She points out that focal points for research could be particular (1) groups or individuals, (2) capabilities, (3) situations or contexts and (4) technological interventions (or a combination thereof). The research agenda should take up both descriptive and normative issues, investigating the contribution of technology to both justice (distribution of capabilities) and ethics (capability expansion). It should aim for the development of general theories with respect to technology and the capability approach, investigate specific domains (such as ICT4D or technology in health care), and conduct case studies. Another job is the identification and specification of capabilities that technology enables, including instrumental capabilities. What is also needed, according to Johnstone, are methods to evaluate technology from the perspective of the capability approach, both at the micro and macro level. In addition I would propose to also pay attention to developing methods or approaches that designers could adopt in order to incorporate the insights of the capability approach in their work. Designers may also benefit from – which Ratan and Bailur (2007) propose – more ethnographic style research on technology and human capabilities.

Insights and input from philosophy of technology is, so the editors of this volume are convinced, relevant for the execution of such a research agenda, of which only

Table 1.1 ICT4D research questions generated from the capability approach (Copied from Zheng 2009)

CA elements	Research questions
Means and ends of development	What kind of "development" is ICTs supposed to promote?
	How do ICTs help people to achieve what they consider to be valuable?
Commodities, capabilities and human diversity	What capabilities can potentially be generated from a certain type of ICT?
	Are they appropriate for local conditions at this stage?
	What conversion factors (personal, social, environmental) need to be in place for capabilities to be generated from a certain type of ICT?
	What decision mechanism affects the actual adoption of a certain type of ICT, or the selection of certain characteristics of a type of ICT over other characteristics?
	How does ICT interact with these decision mechanisms (and their changes)?
Agency and restricted agency	What are the needs and aspirations of the potential ICT adopters?
	What are the rationales behind those needs and aspirations?
	What conditions enable or restrict the 'agency' of the ICT adopters?
	How does ICT interact with these conditions?
Evaluative spaces	What essential capabilities are deprived?
	Who may be disadvantaged by the deprivation of these capabilities?
	What are the relationships between different types of capability deprivations?

Table 1.2 Examples of possible capability-related research questions concerning computer games (Copied from Johnstone 2007)

Direct capability questions	Indirect capability questions
Do computer games improve children's skills or knowledge (enhanced instrumental and substantive freedoms)?	Has the prevalence of computer games reduced the cultural richness of children's environments (diminished substantive freedom)?
Do games reduce opportunities for interaction and play (diminished substantive freedom)?	Have games increased playground pressure to buy expensive consumer products (diminished instrumental freedom)?
What features of games appear to be implicated in enhancing or diminishing specific capabilities?	What contextual factors influence the positive or negative effects that games have on children's environments?
Could games be improved so as to support capability expansion?	

some rough outlines have been sketched here. This discipline may help in conceptualizing how exactly technology fits into or relates to the framework developed by the capability approach. For example, as Kullman and Lee ask in Chap. 3, does technology provide people 'liberation from' or 'liberation within'? And, as

Coeckelbergh addresses in Chap. 5, are technologies only instrumentally important in the creation of universally valid and timeless valued capabilities, or do they also change the meaning of capabilities? Have we – in evaluating a technology or technology-oriented projects – taken sufficiently in account that a technology may have a complex effect on human capabilities, both enabling and constraining them in direct and indirect ways? Is it tenable (Oosterlaken 2009b) that the capability approach fits in with political liberalism, if the technologies needed to create so many human capabilities may not – as many philosophers of technology hold – be neutral towards the good life?

Another important contribution of philosophy of technology – or at least of the work in this field after the 'empirical turn' – may be to point out that rather than talking about technology in general, we need to discuss how concrete technical artefacts, with their specific characteristics and details of design, contribute to the expansion of human capabilities. One of the advantages of adopting a capability approach of technology is that it puts humans at the center of our critical evaluations, as Zheng and Stahl also point out in Chap. 4. Similarly, Gigler (2008) says about his ICT4D evaluation framework based on the capability approach that it

> places, in contrast to the current discourse around the digital divide, the human development of the poor and not technology at the centre of analysis.

And he says that

> a key recommendation of the chapter is that the human development of people, rather than technology itself, should be the center of the design and evaluation of ICT programs.

The capability approach is thus, he claims, to be preferred over more conventional approaches "that overemphasize the significance of technology itself for social change." I fully agree with him. Yet at the same time I think that we should also watch for underemphasizing the significance of the technology itself. For the expansion of human agency and human capabilities it matters, so contemporary philosophy of technology teaches us, what technology exactly we are talking about and what the details of its design are. We should probe into this, without ever letting out of sight that what we are ultimately interested in is not creating ever more advanced gadgets, but making sure that people are empowered to – as Sen would put it – live the lives they have reason to value.

Acknowledgements This research has been made possible by a grant from NWO, the Netherlands Organization for Scientific Research.

References

Ahmed, M. A. (2011). ICTs for information capabilities of parliamentary stakeholders. In Z. Sobaci (Ed.), *E-parliament and ICT-based legislation: Concept, experiences and lessons*. Hershey: IGI Global.

Alampay, E. (2006a). Beyond access to ICTs: Measuring capabilities in the information society. *International Journal of Education and Development Using Information and Communication Technology, 2*(3), 4–22.

Alampay, E. M. (2006b). The capability approach and access to information and communication technologies. In M. Minogue & L. Carino (Eds.), *Regulatory governance in developing countries*. Cheltenham: Edward Elgar.

Alkire, S. (2005a). Why the capability approach? *Journal of Human Development, 6*(1), 115–133.

Alkire, S. (2005b). *Capability and functionings: Definition and justification*. Briefing Notes (last updated September 1, 2005), Human Development and Capability Association. www.hd-ca.org

Alkire, S. (2008). Using the capability approach: Prospective and evaluative analyses. In F. Comim, M. Qizilbash, & S. Alkire (Eds.), *The capability approach; concepts, measures and applications*. Cambridge: Cambridge University Press.

Asveld, L., & Roeser, S. (Eds.). (2009). *The ethics of technological risk*. London: Earthscan.

Barja, G., & Gigler, B.-S. (2007). The concept of information poverty and how to measure it in the Latin American context. In H. Galperin & J. Marsical (Eds.), *Digital poverty: Perspectives from Latin America and the Caribbean*. Ottawa: IDRC.

Birdsall, W. F. (2011). Human capabilities and information and communication technology: The communicative connection. *Ethics and Information Technology, 13*(2), 93–106.

Borenstein, J., & Pearson, Y. (2010). Robot caregivers: Harbingers of expanded freedom for all? *Ethics and Information Technology, 12*(4), 277–288.

Clague, J. (2006). Patent injustice; applying Sen's capability approach to biotechnologies. In S. Deneulin, M. Nebel, & N. Sagovsky (Eds.), *Transforming unjust structures: The capability approach*. Dordrecht: Springer.

Coeckelbergh, M. (2009). Health care, capabilities and AI assistive technologies. *Ethical Theory and Moral Practice, 13*(2), 181–190.

Coeckelbergh, M. (2011). Human development or human enhancement? A methodological reflection on capabilities and the evaluation of information technologies. *Ethics and Information Technology, 13*(2), 81–92.

Comim, F., Qizilbash, M., & Alkire, S. (Eds.). (2008). *The capability approach: Concepts, measures and applications*. Cambridge: Cambridge University Press.

Cooke, E. F. (2003). Germ-line engineering, freedom, and future generations. *Bioethics, 17*(1), 32–58.

Cozzens, S. E., Gatchair, S., et al. (2007). Knowledge and development. In E. J. Hackett, O. Amsterdamska, M. Lynch, & J. Wajcman (Eds.), *The handbook of science and technology studies*. Cambridge: MIT Press.

Crocker, D. A. (2008). *Ethics of global development: Agency, capability, and deliberative democracy*. Cambridge: Cambridge University Press.

Deneulin, S. (2002). Perfectionism, paternalism and liberalism in Sen and Nussbaum's capability approach. *Review of Political Economy, 14*(4), 497–518.

Deneulin, S., & Shahani, L. (Eds.). (2009). *An introduction to the human development and capability approach*. Ottawa: Earthscan/IDRC.

Dong, A. (2008). The policy of design: A capabilities approach. *Design Issues, 24*(4), 76–87.

Foster, J. E., & Handy, C. (2008). *External capabilities* (OPHI Working Paper Series No. 08). Oxford: Oxford Poverty & Human Development Initiative (OPHI).

Frediani, A. A. (2007). Amartya Sen, the World Bank, and the redress of urban poverty: A Brazilian case study. *Journal of Human Development, 8*(1), 133–152.

Garai, A., & Shadrach, B. (2006). *Taking ICT to every Indian village; opportunities and challenges (a collection of four papers)*. New Delhi: OneWorld South Asia.

Gardoni, P., & Murphy, C. (2008). Recovery from natural and man-made disasters as capabilities restoration and enhancement. *International Journal of Sustainable Development and Planning, 3*(4), 317–333.

Garnham, N. (1997). Amartya Sen's "capability" approach to the evaluation of welfare: Its application to communications. *Javnost – The Public, 4*(4), 25–34.

Garnham, N. (2000). Amartya Sen's "capabilities" approach to the evaluation of welfare and its applications to communications. In B. Cammaerts & J. C. Burgelmans (Eds.), *Beyond competition: Broadening the scope of telecommunications policy* (pp. 25–37). Brussels: VUB University Press.

Gigler, B. -S. (2004). *Including the excluded – Can ICTs empower poor communities? Towards an alternative evaluation framework based on the capability approach*. Paper presented at the 2004 conference of the Human Development and Capability Association, Pavia, Italy.

Gigler, B.-S. (2008). Enacting and interpreting technology—From usage to well-being: Experiences of indigenous peoples with ICTs. In C. Van Slyke (Ed.), *Information communication technologies: Concepts, methodologies, tools, and applications*. Hershey: IGI Global.

Grunfeld, H., Hak, S., et al. (2011). Understanding benefits realisation of iREACH from a capability approach perspective. *Ethics and Information Technology, 13*(2), 151–172.

Hamel, J. -Y. (2010). *ICT4D and the human development and capabilities approach: The potentials of information and communication technology*, United Nations Development Program. Human Development Research Paper, 2010/37.

Heeks, R. (2010). Do Information and Communication Technologies (ICTs) contribute to development? *Journal of International Development, 22*, 625–640.

Hellsten, S. K. (2007). From information society to global village of wisdom? The role of ICT in realizing social justice in the developing world. In E. Rooksby & J. Weckert (Eds.), *Information technology and social justice*. Hershey: Information Science Publishing.

Ibrahim, S. (2006). From individual to collective capabilities: The capability approach as a conceptual framework for self-help. *Journal of Human Development, 7*(3), 397–416.

James, J. (2006). The Internet and poverty in developing countries: Welfare economics versus a functionings-based approach. *Futures, 38*(3), 337–349.

Johnstone, J. (2007). Technology as empowerment: A capability approach to computer ethics. *Ethics and Information Technology, 2007*(9), 73–87.

Johnstone, J. (2012). Capabilities. In P. Brey, A. Briggle, & E. Spence (Eds.), *The good life in a technological age*. London: Routledge.

Kleine, D. (2010). ICT4What? – Using the choice framework to operationalise the capability approach to development. *Journal of International Development, 22*(5), 674–692.

Kleine, D. (2011). The capability approach and the 'medium of choice': Steps towards conceptualising information and communication technologies for development. *Ethics and Information Technology, 13*(2), 119–130.

Kroes, P., & Meijers, A. (Eds.). (2000). *The empirical turn in the philosophy of technology*. Research in Philosophy and Technology. Amsterdam: JAI/Elsevier.

Latour, B. (2005). *Reassembling the social: An introduction to actor-network-theory*. Oxford: Oxford University Press.

Madon, S. (2004). Evaluating the developmental impact of E-governance initiatives: An exploratory framework. *Electronic Journal of Information Systems in Developing Countries, 20*(5), 1–13.

Mansell, R. (2001). *New media and the power of networks*. First Dixons Public Lecture and Inaugural Professorial Lecture. London: The London School of Economics and Political Science.

Murphy, C., & Gardoni, P. (2006). The role of society in engineering risk analysis: A capabilities-based approach. *Risk Analysis, 26*(4), 1073–1083.

Murphy, C., & Gardoni, P. (2007). Determining public policy and resource allocation priorities for mitigating natural hazards: A capabilities-based approach. *Science and Engineering Ethics, 13*(4), 489–504.

Murphy, C., & Gardoni, P. (2008). The acceptability and the tolerability of societal risks: A capabilities-based approach. *Science and Engineering Ethics, 14*, 77–92.

Murphy, C., & Gardoni, P. (2010). Assessing capability instead of achieved functionings in risk analysis. *Journal of Risk Research, 13*(2), 137–147.

Musa, P. F. (2006). Making a case for modifying the technology acceptance model to account for limited accessibility in developing countries. *Information Technology for Development, 12*(3), 213–224.

Nussbaum, M. C. (2000). *Women and human development; the capability approach*. New York: Cambridge University Press.

Nussbaum, M. C. (2006). *Frontiers of justice; disability, nationality, species membership*. Cambridge: The Belknap Press of Harvard University Press.

Nussbaum, M. (2008). Human dignity and political entitlements. In *Human dignity and bioethics; essays commissioned by the President's council on bioethics.* Washington DC: The President's Council on Bioethics.

Nussbaum, M. C. (2011a). *Creating capabilities: The human development approach.* Harvard: The Belknap Press of Harvard University Press.

Nussbaum, M. C. (2011b). Perfectionist liberalism and political liberalism. *Philosophy and Public Affairs, 39*(1), 3–45.

Oosterlaken, I. (2009a). Design for development; a capability approach. *Design Issues, 25*(4), 91–102.

Oosterlaken, I. (2009b). *The capability approach, technology and neutrality towards the good life.* Paper presented at the 2009 annual conference of the Human Development and Capability Association, September 9–11, 2011, Lima, Peru.

Oosterlaken, I. (2011). Inserting technology in the relational ontology of Sen's capability approach. *Journal of Human Development and Capabilities, 12*(3), 425–432.

Oosterlaken, I., & van den Hoven, J. (2011). Editorial: ICT and the capability approach. *Ethics and Information Technology, 13*(2), 65–67.

Qureshi, S. (2010). Editorial – Extending human capabilities through information technology applications and infrastructures. *Information Technology for Development, 16*(1), 1–3.

Ratan, A. L., & Bailur, S. (2007). Welfare, agency and "ICT for Development". *ICTD 2007 – Proceedings of the 2nd IEEE/ACM international conference on Information and Communication Technologies and Development,* Bangalore, India. 15–16 December 2007.

Robeyns, I. (2005). The capability approach – A theoretical survey. *Journal of Human Development, 6*(1), 94–114.

Robeyns, I. (2006). The capability approach in practice. *Journal of Political Philosophy, 14*(3), 351–376.

Rubbo, A. (2010). Towards equality, social inclusion and human development in design education: The case of global studio 2005–2008. *Architectural Theory Review, 15*(1), 61–87.

Sen, A. (1983). Poor, relatively speaking. *Oxford Economic Papers (New Series), 35*(2), 153–169.

Sen, A. (1985). *Commodities and capabilities.* Amsterdam/New York: North-Holland.

Sen, A. (1987). Reply. In G. Hawthorn (Ed.), *The standard of living.* Cambridge: Cambridge University Press.

Sen, A. (1990). Capability and well-being. In M. Nussbaum & A. Sen (Eds.), *The quality of life.* Oxford: Oxford University Press.

Sen, A. (1992). *Inequality reexamined.* Cambridge: Harvard University Press.

Sen, A. (1999). *Development as freedom.* New York: Anchor Books.

Sen, A. (2004). Elements of a theory of human rights. *Philosophy and Public Affairs, 32*(4), 315–356.

Sen, A. (2005). Human rights and capabilities. *Journal of Human Development, 6*(2), 151–166.

Sen, A. (2009). *The idea of justice.* Cambridge: The Belknap Press of Harvard University Press.

Sen, A. (2010). The mobile and the world. *Information Technologies and International Development, 6*(special issue 2010), 1–3.

Smith, M. L., & Seward, C. (2009). The Relational Ontology of Amartya Sen's Capability Approach: Incorporating Social and Individual Causes. *Journal of Human Development and Capabilities, 10*(2), 213–235.

Stewart, F. (2005). Groups and capabilities. *Journal of Human Development, 6*(2), 185–204.

Thomas, J. J., & Parayil, G. (2008). Bridging the social and digital divides in Andhra Pradesh and Kerala: A capabilities approach. *Development and Change, 39*(3), 409–435.

Toboso, M. (2011). Rethinking disability in Amartya Sen's approach: ICT and equality of opportunity. *Ethics and Information Technology, 13*(2), 107–118.

van den Hoven, J. (2007). ICT and value sensitive design. In P. Goujon, S. Lavelle, P. Duquenoy, K. Kimppa, & V. Laurent (Eds.), *The information society: Innovations, legitimacy, ethics and democracy* (Vol. 233, pp. 67–72). Boston: Springer.

van den Hoven, J., & Rooksby, E. (2008). Distributive justice and the value of information: A (broadly) Rawlsian approach. In J. van den Hoven & J. Weckert (Eds.), *Information technology and moral philosophy.* Cambridge: Cambridge University Press.

Vaughan, D. (2011). The importance of capabilities in the sustainability of information and communications technology programs: The case of Remote Indigenous Australian Communities. *Ethics and Information Technology, 13*(2), 131–150.
von Tunzelmann, N., & Wang, Q. (2007). Capabilities and production theory. *Structural Change and Economic Dynamics, 2007*(18), 192–211.
Walsham, G., & Sahay, S. (2006). Research on information systems in developing countries: Current landscape and future prospects. *Information Technology for Development, 12*(1), 7–24.
Wresch, W. (2007). 500 Million missing Web sites: Amartya Sen's capability approach and measures of technological deprivation in developing countries. In E. Rooksby & J. Weckert (Eds.), *Information technology and social justice*. Hershey: Information Science Publishing.
Wresch, W. (2009). Progress on the global digital divide: An ethical perspective based on Amartya Sen's capabilities model. *Ethics and Information Technology, 2009*(11), 255–263.
Zheng, Y. (2007). *Exploring the value of the capability approach for E-development*. Paper presented at the 9th international conference on Social Implications of Computers in Developing Countries, Sao Paulo, Brazil.
Zheng, Y. (2009). Different spaces for e-development: What can we learn from the capability approach. *Information Technology for Development, 15*(2), 66–82.
Zheng, Y., & Stahl, B. C. (2011). Technology, capabilities and critical perspectives: What can critical theory contribute to Sen's capability approach? *Ethics and Information Technology, 13*(2), 69–80.
Zheng, Y., & Walsham, G. (2008). Inequality of what? Social exclusion in the e-society as capability deprivation. *Information Technology and People, 21*(3), 222–243.

Chapter 2
Human Capabilities and Technology

Jeroen van den Hoven

2.1 Introduction

Amartya Sen has argued in *On Ethics and Economics* (1987) that economics can benefit from a closer relationship with ethics. I argue analogously that engineering can benefit from a closer relationship with ethics. More in particular, I believe that the capability approach is a normative approach that is highly suitable for bringing ethics and engineering more closely together.

In this paper I do not take issue with comparisons of the capability approach with other answers that have been given to the "equality of what" question, such as resource-based, utilitarian or primary goods accounts, nor do I address meta-ethical issues concerning subjectivism, objectivism, realism or relativism. Nor am I concerned with the way the capability approach should be situated in the landscape of general political philosophical theories. I'll also have nothing to say about measurements of well-being and metrics of equality and justice.

My topic is a fairly limited one, although I think it is of wider importance and needs to be addressed in order for the capability approach to realize its full potential as a practically relevant ethical theory concerned with improving the fate of the global poor and the quality of life in the century of high technology. My paper is concerned with an aspect that was until recently largely and remarkably absent from the literature on the capability approach: technology. I claim that there are close, but unexplored, relations between technology and the capability approach. I refer to this idea as the Capability-Technology-Affinity Thesis (CTA).

After a general characterization of technology in this context (Sect. 2.1) I provide three types of considerations in support of the CTA-thesis. Firstly, we are witness to

J. van den Hoven (✉)
Philosophy Section, Delft University of Technology, P.O. Box 5015, 2600 GA Delft,
The Netherlands
e-mail: m.j.vandenhoven@tudelft.nl

a convergence of two historical developments; a value turn in design and a design turn in ethics (Sect. 2.2). Secondly, a philosophical anthropological observation – based on Ortega y Gasset's analysis of technology and the human condition – about the good life as the terminus *ad quem* of technology (Sect. 2.3). Thirdly, a conceptual analysis of both the capability approach and engineering and of technology as an amplifier of agency and human capabilities (Sect. 2.4).

2.2 Technology in a Modern World

Technology is ubiquitous and all-pervasive. We often forget that nearly everything around us today is artificial and the product of design. We therefore have difficulty appreciating to what extent technology and design contribute to the shaping of our world and of human lives. This seems especially true for moral philosophy, which largely skims over the surface of a world that is deeply technological. Moral philosophers easily overlook the fact that the things we do to each other in this age – things both good and bad – we do directly and indirectly to each other by means of the outcomes of design: instruments, implements, devices and artefacts, organizations, regulation and procedures: from the syringe to the email message and more generally systems of health care, transportation, weapons, energy supply, sanitation, and communication. But modern moral philosophy neglects technology at its own detriment. It may pay the price of becoming largely irrelevant to human lives and societies, which are in important ways shaped and co-constituted by technology.

Even the vital functions of our democratic systems – an important topic for moral philosophers – are in important ways supported and shaped by technology. To give a simple example: The Internet played a crucial role in Barrack Obama's campaign in 2008. What television does for democracies is evident. The design of the so called "butterfly" ballot forms in the 2004 US presidential elections, led to many errors in individual votes and to thousands of invalid forms, which in turn tipped the balance in favor of one presidential candidate, with epochal consequences. US citizens at the time had a right to vote, they wanted to vote, many actually did vote, but whether they had *the capability to vote* can be questioned since the mechanism through which they voted was crooked, biased or unreliable, depending on which reading of the 2004 election you favor.

Philip Pettit remarks that there is a similarity between democratic institutions and general utility infrastructures: they both need to be designed, maintained and need to be made safe for users. Design and development of such infrastructures are loaded with choices and decisions, which may express the values of some and not of others. Bowker and Star (1999, p. 50) refer in this context to the "quiet victories of infrastructure builders inscribing their politics into the systems". To use the example discussed above; those who are in favour of fair presidential elections in the future are well advised to think about the design of voting machines, the computer code, its certification, security, and the question whether these systems need to be open source or based on proprietary software provided by commercial vendors.

The difference between fair and unfair elections today is fixed in the initial stages of design and development of voting machines.

In short: technology and ethics are closely interrelated and we should study these relations in more detail. This also applies to the capability approach of Nussbaum and Sen, a normative framework that has gained popularity in the area of development and global justice. Our technology and engineering could, on the one hand, be greatly improved if we started to take serious the question of how technology and engineering affect human functioning and human capabilities – as specified in the work of Sen, Nussbaum and others. On the other hand, our ethical thinking about functionings and capabilities, could be greatly improved if we would pay close attention to the ways in which technology is designed, developed, and deployed. The affinity of the CTA–thesis is thus bi-directional. I will now discuss three different arguments for this thesis.

2.3 The Convergence of Engineering Design and Practical Ethics

2.3.1 The Ethical Turn in Engineering Design

Fortunately, there is a growing awareness within design and engineering disciplines that ethics is important. Ethics courses for designers and engineers are compulsory in many universities and engineering ethics is growing as a special field of applied ethics. Viktor Papanek already pioneered the unexplored territory of ethical design (including design for development) in the mid-twentieth century, as did for example Enid Mumford (human factors in socio-technical systems), E.F. Schumacher (appropriate technology), and Pelle Ehn (democratic design).

Papanek encouraged designers to think of solutions for the global poor and developing countries, for disabled, for women, for children, for the excluded. He worked for the UN and UNESCO and in 1969 designed a cheap television set for use in Africa that cost less than $10. It could be produced locally with very limited resources. His idea for a radio in 1962 – without batteries, made from used tin cans, costing less than 10 cents – is another illustration of his dedication to design for development. It was successful in Indonesia and India. He developed an artificial seed pod to develop vegetation in arid areas, made of biodegradable plastic, which could be dropped from airplanes. The biodegradable plastic was absorbed by the surrounding vegetation and turned into a fertilizing agent so that the seeds could grow into plants.

Papanek authored several books, one entitled *Design for the Real World* (1971). According to him industrial designers were acting irresponsibly, because they are concerned only with aesthetics and making money and not with other values and human needs. He even claimed there are hardly any professions more harmful than industrial design and engineering design. Industrial designers, to paraphrase

Papanek's own words, by and large only help to produce things nobody needs, for people who can't afford them, to impress people who don't care. The moral opportunities of this profession, however, are formidable. According to Papanek the first question designers should ask themselves is: "Am I on the side of social good, or will the object that I design be an addition to the catalogue of unnecessary fetish objects?". He was a staunch critic of prominent US American consumer society and agreed with Ralph Nader that the wrong values were built into American cars. This observation was made 40 years before GM made its apologies to the American consumers in December 2008 in a daily newspaper advertisement, admitting that they had worked with the wrong values and that they were wrong to focus so heavily on producing unsustainable cars like SUV's, of which the sales have plummeted after the financial crisis of 2008.

A second illustration of the fact that an ethical dawn in the field of technology and engineering is near is sometimes referred to as *Value Sensitive Design* (e.g. van den Hoven 2007). This movement started in Stanford in the late 1980s and early 1990s of the twentieth century and was pioneered by John Perry, Terry Winograd, and Batya Friedman. They experimented with computer interfaces for disabled. Gradually this group became aware through its practical work of the biases and value assumptions that were often built into technology. And they started to think about Value Sensitive Design (VSD) as a way of developing IT that aims at making the right moral values part of technological design, research and development. It assumes that human values, norms, moral considerations can indeed be imparted to the things we make and use and it construes information technology (and other technologies for that matter) as a formidable force which can be used to make the world a better place, especially when we take the trouble of reflecting on its ethical aspects in advance.

If our moral and political discourse on user autonomy, patient or citizen centeredness, privacy, or security is to be more than an empty promise, these values will have to be expressed in the design, architecture and specifications of systems. If we want our information technology – and the use that is made of it – to be just, fair and safe, we must see to it that it inherits our good intentions. If we think that all people should, as Nussbaum defends, have some valuable human capabilities necessary for people to have dignified human lives, necessary to lead the lives they have reason to value, then we must make this an integral part of our discussions about the details of design. Moreover technologies must be seen to have those properties and we must be able to demonstrate that they possess these morally desirable features, compare different applications from these value perspectives, motivate political choices and justify investments from this perspective.

2.3.2 The Design Turn in Applied Ethics

In the same period that design started to turn slowly towards ethics, ethics went through a converging development. In the mid-twentieth century applied ethics became fashionable. It has gradually become more respectable. Many philosophers now seem to think, as Dewey (1917) did once, that "philosophy recovers itself when it ceases to be a device for dealing with the problems of philosophers and becomes a method, cultivated by philosophers, for dealing with the problems of men". Among them are Hilary Putnam, Richard Posner, Martha Nussbaum, Michele Moody Adams. This is a good starting point, but it is not enough. The 'applied turn' is being pushed just one step further.

Institutional and technological contrivances empower us in a myriad of ways. However, these very same humanly designed institutions and technologies also constrain our freedom to act and to know. Motorways enable us to travel speedily by car from A to B; but motorways also require us to drive on one side of the road and along a predetermined pathway between A and B. Hospitals and pacemakers enable us to save and prolong lives. On the other hand military organizations and weapons of mass destruction allow us to destroy lives. Frequently institutions and technologies, like genetic engineering, have dual uses, ethically speaking: they can be used for good and for evil. So design products empower us and constrain us, they enable us to do certain things and prevent us from doing other things; moreover they are ubiquitous and in part constitutive of our human environments. As such design products have an important normative dimension.

Sometimes a moral end or feature is *designed into* an institution or technology. Sometimes a morally desirable outcome is the fortuitous, but unintended consequence of an institutional arrangement or technological invention. The paradigm case of a morally objectionable end being *designed into an artefact* are Robert Moses' low hanging overpasses in New York, which were intended to prevent the busses from the poor black neighbourhoods to be routed to the beaches near New York – a favourite destination of white middle class families (Winner 1980).

The design turn in applied ethics can be construed as the third and most recent phase in the development of contemporary ethics. After an exclusive focus on meta-ethics in the beginning of the twentieth century, there was an applied turn in the latter decades of the twentieth century. However, this applied turn consisted of the application of existing theory to practical problems and primarily involved the adjudication between predetermined options, often on the basis of weighing up the moral considerations inherent in these given options. The design approach in applied ethics, seeks to create or expose additional options, to reconfigure or re-design the option set. An important additional set of questions needs to be articulated in applied ethics. Not only the question 'given this situation and the fact that the options A and B are open to the individual (doctor in an emergency situation, person in a prisoners' dilemma, person in a trolley case[1]), what should she do?', but also the question 'who designed the situation in such a way that only these two options A and B are open to her?' needs to be addressed. We should ask ourselves how the design of institutional arrangements, incentive structures, infrastructures and artefacts could be improved.

---

[1] The trolley problem is an often discussed thought experiment in ethics, first introduced by Philippa Foot (1967), in which a person is put in a difficult ethical dilemma concerning how to respond to a threatening situation involving an approaching trolley.

Moral philosophers working on real life problems are rising to the occasion and seem to realize that design matters[2]: of institutional arrangements, incentive structures, legal frameworks, business processes, voting procedures, business processes, but also of protocols, computer programs and information systems and a great variety of technical and engineering artefacts and infrastructures. Design products have an important shaping potential in society and in individual human lives. If moral philosophers want to be really useful, they need to make their analyses available when it still matters and when it can be used to inform the design of things to come, both technological and institutional.

2.4 Technology and the Good Life

Apart from the fact that both engineering/design and ethics seems to be independently set on paths that are slowly converging, there is a fundamental reason for bringing them together. They have the same *terminus ad quem*: the good life. The evaluation of our achievements and contributions to both should therefore be made with the same yard stick.

The Spanish philosopher Ortega y Gasset (1972) is one of the few general philosophers of the twentieth century who has addressed questions concerning technology and the good life.[3] According to Ortega it is important to realize that human beings (members of the species *Homo Sapiens*) can in principle do without technology. We would be cold, miserable, hungry and uneasy much of the time without clothing, shelter, fire, cooked food, weapons for self-defense or hunt, grindstones, plow, utensils. This is the state in which other primates find themselves. With the first tools that were introduced, Homo Sapiens tried to improve his situation, every new artefact which is introduced aims at making the world a better place for him to live: making things less tedious, less strenuous, less cold, less painful, easier, more pleasant, easier to digest. Sometimes we were mistaken, inaccurate, self-deceived, incompetent in judging the efficaciousness of our tools, but the aim always is and has been to make things better. The *terminus ad quem* of technology is therefore the good life. This also explains the variety of technologies: so many different conceptions of the good life, so many different technologies and artefacts. According to one of the prominent historians of technology, Basalla (1989), the immense variety in artefacts and technology is directly related to the many different conceptions of the good life. Ortega puts it as follows: "Man's desire to live, to be in the world, is inseparable from his desire to live well". According to Ortega "Man, technology, wellbeing

---

[2] Russell Hardin was one of the prominent contributors to *The Theory of Institutional Design* (Cambridge UP, 1996) and has worked, like Pettit, Brennan, Dryzek, Luban, Goodin and others on institutional design. Institutional, constitutional, legal and governance arrangements are the object of design efforts and they need to be explicitly evaluated and analysed as such. Hardin claimed that morality needs to be *designed in* to the organizational structure. Others who have focussed on design approaches are Pogge, Sunstein and Thompson. See also Pogge et al. (forthcoming).

[3] See van den Hoven (2012) for a detailed account of Ortega's view on technology.

are in the last instance, synonymous. Only when we conceive of them as such are we able to grasp the meaning of technology …"

Ortega makes a further important observation, namely that there is a difference between functioning (e.g. keeping one's body temperature around 37°C) and building a shelter (having the capability to keep one's body temperature around 37°C). The latter is an engineering intervention in the world, which requires a design and planning and range of instrumental actions, including joint actions. We change the world in order to cater for our needs, to accommodate ourselves. We temporarily stand back from our immediate needs to keep warm and may decide to work in the freezing cold on something we hope will keep us warm in the future. We change the world by constructing a shelter that gives us the capability to keep the cold out in the night, and the sun during the day. The technological actions have as their *telos* to keep warm, but are not directly achieving this at the present moment only: "Technology is a reform of nature, such that the satisfaction of necessities is guaranteed under all circumstances".

A last observation that we would like to draw attention to is that Ortega describes how identities and moral autonomy are related to technology. Technology is not only important to achieve certain functionings it may also be necessary to be a particular type of person who is the subject of those functionings. It is impossible to be a Samurai without a special sword that is quite difficult to make, one cannot be an English gentlemen without a good supply of running water and razorblades, one cannot be a young urban professional without a mobile phone and laptop computer.

Every technological device, machine, artefact, piece of infrastructure comes with the implicit or explicit suggestion that we would be better off if we used it (although those who make the suggestion can be – and sometimes definitely are – mistaken, wicked, or confused). Even the inventors of weapons of mass destruction probably thought that their inventions were going to make the world a better place. Technology and infrastructures should therefore always be evaluated in their quality of contributing (or not contributing, as the case may be) to people's capabilities to lead flourishing human lives, since that is what they fundamentally aim at, it is their *terminus ad quem*. Although technologies overall expand people's capabilities to lead the lives they have reason to value, we should at the same time recognise that technologies often have good life implications transcending the level of the individual making choices about his/her own life. Since individuals moreover often disagree about the question what the good life is, it may be too simple to suppose – as the capability approach does – that a policy focus on human capabilities instead of human functionings allows one to remain largely neutral towards the good life (Oosterlaken 2009b).

2.5 The Capability Approach and Technology: Structural Similarities

Human capabilities have been described as the positive freedoms or effective, real opportunities that people have to do and be certain things. These beings and doings are called 'functionings' by Sen. Functionings "together constitute what makes a life valuable" (Robeyns 2005, p. 95) and are "constitutive of a person's being" (Alkire 2005, p. 118). Examples of functionings mentioned in the literature are such diverse things as working, resting, being literate, being healthy, being part of a community, being able to travel and being confident. Some functionings may be very basic (being nourished, literate, clothed) and others might be quite complex (being able to play a virtuoso drum solo). "The distinction between achieved functionings and capabilities", so Robeyns (2005, p. 95) explains, "is between the realized and the effectively possible; in other words, between achievements on the one hand, and freedoms or valuable options from which one can choose on the other". This distinction is not unlike the distinction made by Ortega between functioning (being warm) and building a shelter (having the capability to stay warm).

In the capability approach 'human functionings' thus belong – together with 'capabilities' – to the core concepts. The intimate relation between human capabilities and technology is already indicated by the striking resemblance of terminology used in both fields. It is not a coincidence that, as Sen (1990) himself has observed, the term "capability" has a technocratic sound, which reminds one of its use in expressions such as "nuclear capability". In engineering artefact 'function' is a central concept and 'malfunctioning' of technical artefacts something one would like to avoid. As Johnstone already remarked about the capability approach:

> because the theory is essentially naturalistic and functionalist in orientation, capability analyses are able to integrate descriptive and normative dimensions in a way that is particularly appropriate to technological domains (Johnstone 2007, p. 84)

It is also noticeable that capability theorists have often used a technical artefact, namely a bicycle, to explain the focus of their approach:

> Having a bike gives a person the ability to move about in a certain way that he may not be able to do without the bike. So the transportation *characteristic* of the bike gives the person the *capability* of moving in a certain way. That capability may give the person utility or happiness if he seeks such movement or finds it pleasurable. So there is, as it were, a *sequence* from a commodity (in this case a bike), to characteristics (in this case, transportation), to capability to function (in this case, the ability to move), to utility (in this case, pleasure from moving). (Sen 1983, p. 160)

However, when it is said that "we are not interested in a bicycle *because it is an object made from certain materials with a specific shape and color*" (Robeyns 2005, p. 98), this seems to indicate – as Oosterlaken (2009a) has remarked – naivety amongst capability theorists regarding the sociology and philosophy of technology.

Indicative of the intimate relationship between technology and human capabilities is perhaps also that capability theorists and engineers have developed some similar insights, independently of each other. A central argument in the capability literature is that capabilities are a more suitable space of equality than resources, since all sort of personal, environmental or social conversion factors may hinder the conversion of resources as such into valued human capabilities. Some engineers, independently, have also realized how important such conversion factors are in the case of technological resources (Oosterlaken 2009a). For example, the universal design movement (e.g. Nieusma 2004) propagated taking personal conversion factors already into account in the design itself, so that the technology would 'compensate' for them and be appropriate for widely diverse groups of users, such as the disabled, women, the aged, the infirm, and the young.

To conclude this section, I would like to propose to characterize technical artefacts and devices (levers, machines, tools, implements, instruments) as *agentive amplifiers*. They create possibilities we would not have without them. They may help us to get enough oxygen where we wouldn't otherwise be able to breath (respiratory disease, altitude, under water, pollution, etc), to get nutrients (cooking, logistics, processing) out of the organic environment, to move around (vehicle, artificial limb), to communicate (phone, computer). A simple counterfactual conditional analysis brings to light the structural similarities between human capabilities and technology. The only possible world in which a handicapped person has the capability to do X, is the world where there is a technology available (e.g. prosthesis) which provides the relevant functionality. The person would not be able to function normally if the relevant technology would not have been available.

Technology can make the impossible possible for people, not in the sense of "logically possible," of course, but in the sense of "feasible" or "physically realizable." It allows them to access possible worlds that would have been inaccessible without it. Given technologies S and T, where S is less advanced than T, it may be the case that A is not possible with technology S, but A is possible with technology T. It is possible that A, but only in the presence of T.[4] When we pause to reflect, there is only very little that we can do and be, that we could achieve without either modern technology or more primitive technologies. Interrelations between human capabilities and technologies can, however, be more complex than has been revealed by the before mentioned counterfactual conditional analysis. Technology may contribute to the expansion or reduction of capabilities in a direct or indirect way. Technology may expand capabilities of some groups, while reducing capabilities of some other group. Technology may expand some capabilities in the short term, while reducing other capabilities in the long term (or the other way around). And capability effects may be intended or unintended, expected or unexpected. In short: the capability effects of new technologies may be complicated. Further research is necessary to better understand them.

2.6 Conclusions

In this chapter, I have shown – using historical, philosophical-anthropological and conceptual arguments – that there exist close relationships between engineering and the capability approach. This was only sparsely discussed in the literature until quite recently. It is a gap that needs to be addressed in order for the capability approach to realize its full potential as a practically relevant ethical theory concerned with improving the fate of the global poor and the quality of life in the century of high

---

[4] See for a detailed and formal account van den Hoven et al. (2012).

technology. The relationships between technology and human capabilities can be complex, depending both on the details of design and on the features of the context of application. Further research is necessary, both philosophically and empirically, in order to understand these interrelations and to make optimal usage of technology for creating well-being, justice and equality.

Acknowledgments This research has been made possible by a grant from NWO, the Netherlands Organization for Scientific Research.

References

Alkire, S. (2005). Why the capability approach? *Journal of Human Development, 6*(1), 115–133.
Basalla, G. (1989). The evolution of technology. In G. Basalla & O. Hannaway (Eds.), *Cambridge studies in the history of science*. Cambridge: Cambridge University Press.
Bowker, G. W., & Star, S. L. (1999). *Sorting things out: Classifications and its consequences*. Cambridge: MIT Press.
Dewey, J. (1917). The need for recovery of philosophy. In: *Creative Intelligence: Essays in the Pragmatic Attitude* edited by John Dewey. New York: Holt.
Foot, P. (1967). The problem of abortion and the doctrine of double effect. *Oxford Review, 5*, 5–15.
Johnstone, J. (2007). Technology as empowerment: A capability approach to computer ethics. *Ethics and Information Technology, 9*(1), 73–87.
Nieusma, D. (2004). Alternative design scholarship: Working towards appropriate design. *Design Issues, 20*(3), 13–24.
Oosterlaken, I. (2009a). Design for development; a capability approach. *Design Issues, 25*(4), 91–102.
Oosterlaken, I. (2009b). *The capability approach, technology and neutrality towards the good life*. Paper presented at 2009 annual conference of the Human Development and Capability Association, September 10–12, 2009, Lima, Peru.
Ortega y Gasset, J. (1972). Thoughts on technology (H. Weyl, Trans.). In C. Mitcham & R. Mackey (Eds.), *Philosophy and technology: Readings in the philosophical problems of technology*. New York: Free Press.
Papanek, V. (1971). *Design for the real world: Human ecology and social change*. New York: Pantheon Books.
Pogge T., Miller, S., & van den Hoven, J. (forthcoming). *The design turn in applied ethics*. Cambridge: Cambridge University Press.
Robeyns, I. (2005). The capability approach – A theoretical survey. *Journal of Human Development, 6*(1), 94–114.
Sen, A. (1983). Poor, relatively speaking. *Oxford Economic Papers (New Series), 35*(2), 153–169.
Sen, A. (1987). *On ethics and economics*. Oxford: Basil Blackwell.
Sen, A. (1990). Justice: Means versus freedoms. *Philosophy and Public Affairs, 19*(2), 111–121.
van den Hoven, J. (2007). ICT and value sensitive design. In P. Goujon, S. Lavelle, P. Duquenoy, K. Kimppa, & V. Laurent (Eds.), *The information society: Innovations, legitimacy, ethics and democracy*. Boston: Springer.
van den Hoven, J. (2012). Neutrality and Technology: Ortega Y Gasset on the Good Life. In P. Brey & A. Briggle (Eds.), *The Good Life in a Technological Age*. London: Routledge.
van den Hoven, J., Van de Poel, I., & Lokhorst, G.-J. (2012). Engineering and the problem of moral overload. *Science and Engineering Ethics, 18*(1), 143–155.
Winner, L. (1980). Do artifacts have politics? *Daedalus, 109*(1), 121–136.

Part II
Technology

Chapter 3
Liberation from/Liberation within: Examining One Laptop per Child with Amartya Sen and Bruno Latour

Kim Kullman and Nick Lee

3.1 Introduction

As Ilse Oosterlaken (2009, 2011) argues, the capabilities approach of economist Amartya Sen offers an inspiring setting for exploring thematic and theoretical overlaps between the worlds of design and development studies. Sen, as is well known, claims that economic evaluations of human wellbeing often overlook the real life conditions of developing nations. He therefore draws our attention to the varied circumstances where persons 'convert' goods and services into actual 'capabilities' that enable them to shape their own lives. In doing so, Sen hints at the intimate connection between design and human value: he invites us to consider ways of realizing and evaluating technological projects that are more sensitive to human diversity.

Over the past two decades, a comparable approach to the relations between technologies, capabilities and human value has taken shape in the work of sociologist Bruno Latour and the substantial literature produced by the anthropologists, geographers, psychologists and sociologists he has influenced. Across a series of ethnographies on topics in science and technology, such as laboratory practices (Latour 1988), transportation design (Latour 1996) and urban infrastructure (Latour and Hermant 1998), Latour has elaborated methods and concepts that

---

K. Kullman (✉)
Department of Social Research, University of Helsinki, P.O. Box 18,
00014 Helsinki, Finland
e-mail: Kim.Kullman@helsinki.fi

N. Lee
Institute of Education, University of Warwick, Coventry CV4 7AL, UK
e-mail: N.M.Lee@warwick.ac.uk

are helpful in tracing the ongoing and mutual shaping of society, technology and nature. Not unlike Sen, Latour seeks to understand human agency and diversity through exploring our actual involvement in everyday social and material environments.

The objective of our chapter, then, is to stage an encounter between Sen and Latour. We will explore differences and similarities between the two authors, and we will ask whether Latour's approach could be used to ground Sen's ideas of development and human value as phenomena open to empirical study. We will argue that despite their contrasting disciplinary and theoretical backgrounds there are plenty of productive parallels between Sen and Latour. Both of these authors are, in their own distinct ways, invested in a vision where more lasting capabilities and freedoms are not the result of unmaking relations between persons and material environments but rather emerge through a careful reordering of those relations.

More precisely, we argue, first, that overcoming the tensions between Sen and Latour opens a fresh passage between development studies and science and technology studies. Combining the two areas of inquiry enhances our ability to explore how notions of human value are folded into technologies designed for citizens of developing nations and how these technologies in turn shape and become shaped by actual users. Understanding Sen and Latour together also provides critical purchase on the 'biopolitical' implications of such designs, that is, their possible role in contests for use of the capabilities and potentials of majority world populations (see Bull 2007; Rose 2007). We think that there is every reason to believe that promises of a technological fix to problems of both over-development and under-development are becoming increasingly appealing in global marketing terms.

Second, combining Latour and Sen broadens the possibilities for evaluating whether abstract notions such as 'freedom' and 'development' are actually realized in encounters between new technologies and citizens of developing nations. Where 'freedom' is often understood as liberation *from* one's social and material surroundings, we will use the Sen/Latour encounter to clarify a view of freedom as liberation *within* one's environment. Liberation does not involve rendering the environment irrelevant by breaking existing relations with it. Instead liberation is composed through gradual changes in everyday socio-technical relations of specific collectives. For the task of connecting design and development, this view extends to those relations established within technological artefacts in the course of their design and distribution.

Our methodological concern with 'relationships' in the production of liberation and freedom will be made concrete through an examination of the One Laptop per Child (OLPC) program, which seeks to bridge global knowledge and digital divides by providing children of developing countries access to low-cost and energy efficient laptop computers. We will illustrate the above 'from/within' contrast and its implications for development through technology and design with a critical discussion of the promotional materials generated within the OLPC and the limited evidence that currently exists about the actual use of OLPC by children of developing countries. Leaving momentarily aside the question of its success as a program, OLPC connects design issues with matters of social justice on a global scale. At the same time, the program engages with cultural assumptions about the 'development' of children by seeking to build capacities for them through the shaping of technology (see Lee 2005). Before exploring OLPC more closely, however, we will introduce Sen and Latour as well as note some of their theoretical similarities and differences.

3.2 Sen, Latour and the Politics of Knowledge

Sen and Latour have long and varied publishing careers and have each inspired research traditions (see Law and Hassard 1999; Morris 2010). Our aim in this section is not to provide an exhaustive account of their work, nor is it to assess the value of the traditions they have spawned. Instead we will discuss the theoretical maneuvers that Sen and Latour make within the 'politics of knowledge' regarding relations between humans and technologies as well as agency and liberation. Although the two have different starting points and are involved in discrete struggles, we argue that they share some core commitments.

The first thing to note is that Sen and Latour formulate their arguments in distinct settings. Sen (1999, 2010) is primarily concerned with the wellbeing of the populations of the developing world. These populations are considered both as an object of government and as potentially autonomous agents. For Sen, a key purpose of government is to enhance the capacities for self-determination of populations, especially where inequality and injustice have historically diminished them. Latour (1993, 2004), in contrast, writes from within the developed world, where 'development' is seen as rooted in technological 'progress'. Technologies gain their economic and political function when they generate new products for the market, increase the efficiency of production or give rise to new media of exchange. According to Latour, development through technology has certainly contributed to the wealth and power of the minority world, but this success has also reinforced the problematic view that humans and their environments are fundamentally distinct, and that progress therefore entails greater separation between them.

At this point, the differences are relatively clear: Sen is concerned with the means by which self-determination can be fostered in populations who have experienced political and economic oppression, while Latour is interested in the self-comprehension of the developed world in its relation to technology and nature. However, we have also suggested that the work of these two authors can be understood in terms of the 'politics of knowledge'. What we mean by this will become clear as we now turn to describe in closer detail the different problems that Sen and Latour identify and address.

3.2.1 Sen's Politics of Knowledge

For Sen, the task of governments is to encourage the self-determination of their populations by enabling people to make decisions for themselves and shape their own lives—this, he argues, is 'liberation' (Sen 1999, 2010). Major obstacles to liberation include poverty as well as economic and political inequality. Poverty limits the ability of individuals to approach their circumstances strategically by planning for the future. Inequalities tend to further diminish self-determination since those unable to act strategically are forced to put themselves in the service of others, sometimes in order simply to survive. Where governments and other powerful actors are inclined to act against poverty and inequality to promote self-determination, the design of practical interventions requires that poverty and inequality are defined in concrete terms. It is at this point that the work of Sen engages with the politics of economic knowledge. According to him, the intellectual tools that are most frequently used to convert abstract concerns with poverty and inequality into practical interventions are often unsuited to the task.

One way to measure poverty and inequality is to evaluate people's access to 'commodities'—useful or tradable items. On this view, roughly speaking, the greater the quantity and range of commodities someone has access to the less poor they are. Likewise, the more two individuals have access to the same number and range of commodities, the more equal they are. Were this mode of measurement a full description of poverty and inequality, then anyone inclined to reduce them should try to ensure that greater ownership of useful or tradable items is extended to the poor. All things being equal this view makes good sense. For Sen (1985), however, measuring commodity ownership can be an insufficient proxy for evaluating poverty, which is why such policies will not necessarily foster self-determination. Dominant economic knowledges ignore the need to 'convert' commodities into the 'capabilities' that lie at the heart of agency. Recognizing this requires more attention to human diversity and the concrete circumstances where people live than the 'commodity' view allows. Two individuals who own the same commodity may well have different abilities to mobilize or trade that commodity so as to adapt it for personal needs—for instance, certain types of food may be unsuitable for bodies weakened by malnutrition and some technological appliances may be useless without electricity infrastructures (see Sen 1983, 1985, 1999).

As the notions of 'conversion' and 'capability' will be discussed in more detail below, let us simply reiterate here that Sen's distinction between commodities and capabilities constitutes an attempt to intervene in the politics of economic knowledge so as to increase the possibility of self-determination and liberation amongst developing world populations. Sen shows that capabilities can be expressed within economic theory and therefore be used to guide decision-making. Unless more adequate understandings of poverty and inequality are formulated and used to shape governmental decision-making, interventions against them are likely to have limited or perverse effects. Even worse, a history of failures could discourage future interventions.

3.2.2 Latour's Politics of Knowledge

Latour also engages in the politics of knowledge. Nevertheless, whereas Sen challenges key economic accounting practices, Latour questions dominant assumptions about technological development. For Latour (1993), the organizations and discourses that promote and manage technological development in Europe and America are strongly influenced by the view that there is a clear ontological distinction between the realms of the human and the non-human, between 'society' and 'nature', 'humans' and 'things'. Politics and science are therefore seen as separate arenas of endeavor, and mixing them becomes dangerous.

According to this prevalent idea, modernity and development are trends toward the greater separation of society from nature and the increasing transcendence by humans of the non-human world. This increase of human powers and liberation from dependency is seen as a result of technological progress. However, rather than simply refuting such a view, Latour takes a step back to examine its conditions of possibility. He notes that a great deal of decision-making, planning and debate about science and politics is predicated on this purifying separation of the human and the non-human. Yet the sense of independence and liberation of the human subject has only been made possible through an increasingly intimate and widespread involvement of humans with non-human elements, from the molecules of chemistry to the geological materials of road infrastructures. As Latour reminds us time and again, without this involvement human societies and actions would neither hold together nor be able to reach across spatial and temporal distances.

Even as our accounts of human liberation, progress and technology have spoken more about human distinctiveness, the ontological separation between human and non-human worlds is constantly undermined in practice. Any form of action always depends on the successful formation of 'hybrids'—mutually strengthening relationships of materials, be they nominally human or non-human (Latour 1993). Latour (1988, 1999) often describes this ceaseless folding together of human and non-human materials with the notion of 'translation', which may be taken as a companion concept to Sen's 'conversion' and which we will explore more closely in subsequent sections. For the moment, it suffices to say that translation highlights the ongoing and uncertain adaptation work between society, technology and nature that is required for any kind of human action and association to be possible.

Attending to processes of translation, then, is essential for understanding current societies. Latour (2004) illustrates this by directing his attention to two major contemporary issues of development—climate change and bio-technological innovation, such as genetically modified organisms in farming. Addressing the problems and potentials of each requires that we think and act across the apparent divide between human and non-human worlds, politics and science. However, the Euro-American commitment to the general ontological distinction between society and nature, humans and things ensures that these activities exceed both the grasp of government planning and the terms of democratic decision-making processes. Latour's language of 'actor-networks', 'collectives', 'hybridity' and 'translation' are therefore his attempt to build knowledges appropriate to the challenges posed by these future developments (see Latour 1999, 2004, 2005).

3.2.3 Sen and Latour Considered Together

At first glance, our sketches indicate a stark contrast between Sen and Latour. Sen seeks to calibrate interventions in the pursuit of self-determination and liberation for human populations, whereas Latour questions the pertinence of 'liberation' to understanding human agency in current societies. Sen's concerns are broadly humanist, while Latour warns that a version of humanism that emphasizes our distinctiveness is impeding responses to global challenges. The different settings where Sen and Latour work and their specific strategies in the politics of knowledge seem to set them at odds, so as to stymie any attempt to combine their approaches.

From our point of view, however, both Sen and Latour have much to contribute to the analysis of contemporary issues in technological design and development. As we will indicate in the following sections, there is a clear structural resemblance between Sen's moment of 'conversion' of commodity into capability and Latour's moment of 'translation' of human and non-human materials into situated forms of agency and association. We have already seen that Sen asserts the specificity of encounters between persons and commodities against the utility function. On his view, no general principles will be adequate to the task of delivering social justice in practice: liberation is possible, but grows from embedded encounters. Likewise, Latour asserts the specificity of encounters between nominally human and non-human materials against an ontology in which they are understood as different kinds of entity. Viewing humans and non-humans as belonging to distinct realms gives rise to a distorted reflection of the source of human agency. For Latour, human development, where it can be said to happen, is not liberation from non-human materials but rather takes place in collectives of human and non-human elements.

At the risk of taking liberties with the carefully crafted strategies of Sen and Latour, we can now sharpen the similarities in their politics of knowledge. We may formulate the question about technics and liberation in at least two ways:

> How can I transcend the surrounding social and material relations or enable others to do this so that it becomes safe to ignore those relations?

> How are the surrounding social and material relations that define me composed and what opportunities do I have to alter them and thus myself?

The first question is based on the concern to liberate *from*. It seeks to establish zones of 'pure' autonomy for persons so that they are able to distance themselves from their immediate environments. On this view, liberation is a process whereby a person shifts from a state of dependence to a state of independence, from being bound to being separate. The second question is driven by a concern to liberate *within*. It does not envisage circumstances where one is separate from the environment, because relationships are vital for agency, whether individual or collective. It suggests that improving human conditions and capabilities does not necessarily require heroic measures of transcendent security but rather a careful and gradual 'conversion' or 'translation' of everyday relationships in order to make more space for situated forms of human 'development' and 'growth'. We would like to argue that both Sen and Latour are more compatible in their views on humans and technologies, agency and liberation with the second formulation. Having established our key distinction (liberation from/within) and put forward an argument that aligns Sen and Latour despite their differences, we can now turn to consider our empirical case.

3.3 OLPC: First Empirical Interlude

One Laptop per Child was founded in 2005 at the Media Lab of Massachusetts Institute of Technology by its director Nicholas Negroponte. It is a non-profit project for the design, manufacture and distribution of low-cost, wireless-connected XO laptop computers. The project is committed to giving the world's poorest children access to education and is based on the view that digital processing technology can empower individuals by fostering communication. At each stage of the project, decisions have been made that aim at fitting the laptops to the core purpose of giving each child the opportunity for 'self-empowered learning.'[1] There have been a number of incarnations of the device and a multi-touch tablet version is in development, but each has been designed to be suitable for child users and to cope with dust, high heat and humidity. Every computer has a screen that stays visible in sunlight, two earlike antennae, a colorful body and a QWERTY keyboard. The use of flash memory in preference to a hard disk makes the device robust and energy efficient, and some versions of the machine include solar and hand-cranked power source options. The original marketing model had the laptop priced low enough ($100 was the target) for governments of developing nations to acquire in large numbers.

The OLPC program is inspired by ideas from constructionist learning theory, which lays emphasis on the inquiry-based and self-directed learning of children rather than on formal approaches to education that build on curricula and instruction by adults. The pioneering work of computer scientist and educational theorist Seymour Papert (1980) on IT as a tool for learning in classrooms has been an important inspiration for OLPC. Another central influence on the OLPC program is the founder Nicholas Negroponte, whose research within the field of computer science has examined the democratic and liberatory potential of the development and diffusion of digital technology—a topic that is expressly elaborated in Negroponte's bestselling book *Being Digital* (1995).[2] Most of these ideas are reflected in the five key principles of the OLPC program:

1. Each child gets an individual laptop
2. Laptops are designed for early education (6–12-year-olds)

---

[1] http://laptop.org/en/vision/

[2] http://laptop.org/en/vision/project/index.shtml

3. Laptops are distributed evenly among pupils
4. Laptops are connected to the internet
5. Laptops use free and open source software[3]

Giving children individual laptops is seen as enabling them to move beyond the limitations that characterize educational settings in developing nations, such as the lack of electricity networks, internet access and school facilities. The adaptable and wireless features of the laptops as well as the use of open source software are expected to open an affordable shortcut for children in developing nations to connect with the same learning spaces as Western children, thereby bridging global knowledge and digital divides. Such goals are also supported by the even distribution of laptops among children in order to increase the 'digital saturation'.[4] These hopes are further crystallized in the OLPC promotional video, which stresses that using the laptop as an educational tool invites children to 'teach themselves', each other, their parents and, some day, the world. "Give a laptop—change the world."[5]

After being announced in 2005 at the World Economic Forum in Davos, Switzerland, the OLPC program has attracted broad international media attention as well as sponsors and industry collaborators such as Quanta, a Taiwan-based company manufacturing hardware for the laptop, eBay, Google, News Corporation and Microsoft. The first XO laptops became available in 2007 and altogether some 1.85 million devices were "delivered, shipped or ordered" across the Americas, Asia and Africa by mid-2010.[6] This is still a major reduction from the original target of 100–150 million devices distributed by the end of 2007 (Kraemer et al. 2009).[7]

Although the intention of this chapter is not to argue for or against the OLPC project—a task it leaves for more thoroughgoing empirical studies and evaluations—it is worthwhile to review some of the noted shortcomings of the program. The first problem that previous research raises is the relatively high acquisition cost of the laptops (e.g. James 2010; Kraemer et al. 2009), which in 2010 was around $199. Adding deployment, maintenance and teacher training expenses to this sum means that buying laptops may easily lead to resource imbalances in the already strained educational budgets of developing nations (James 2010).

---

[3] http://wiki.laptop.org/go/OLPC:Five_principles

[4] http://wiki.laptop.org/go/OLPC:Five_principles

[5] http://laptop.org/en/vision/index.shtml

[6] In 2007 and 2008, OLPC organized 'Give One Get One' campaigns that enabled private persons in the U.S. to donate one laptop and receive one for themselves for $399. The first campaign was successful (167,000 units sold), whereas the second resulted in markedly fewer donations (12,500 units sold) (see Kraemer et al. 2009). Currently, it is possible for private persons to donate XO laptops at the cost of $199 apiece via the Amazon website www.amazon.com

[7] Most of the OLPC programs across the world are at a pilot stage with governments ordering between 1,000 and 10,000 units at a time. Countries that have participated actively include Uruguay (420,000 units), where every primary school pupil now has a personal XO laptop, Peru (290,000 units), Rwanda (110,000 units) Argentina (60,000 units) and Mexico (50,000 units).This estimate is taken from the OLPC wiki pages (http://wiki.laptop.org/go/Deployments). Others have noted that exact figures are difficult to acquire (see James 2010; Kraemer et al. 2009).

The second problem of the OLPC project has to do with the deployment of the computers. Despite an attempt to design a low-cost device adaptable to various settings, the implementation of the program has been complicated by the shifting cultural, economic and political characteristics of the environments where OLPC operates. Apart from having to deal with lacking means and infrastructures, OLPC has found it challenging to adapt its approach to local decision-making processes and institutional arrangements (see Kraemer et al. 2009).

The third issue concerns the commercial aspect of OLPC. XO laptops have encountered fierce competition from PC manufacturers, who regard developing nations as an emerging market for their products. Intel, a company that briefly collaborated with OLPC but withdrew due to disagreements with Negroponte, has developed an alternative to the XO laptop, called Classmate. Although its average price of $250 is relatively high compared to XO, Intel has managed to enter the education market in developing nations through laptop donations and free teacher training (Kraemer et al. 2009). Whether OLPC will be able to respond to the challenge posed by the growing need for low-cost laptops is an unresolved question, but recent collaboration with Microsoft points to a change in its approach. XO laptop buyers can now add a Windows XP operating system and Microsoft Office to the device at extra cost. This has led some to ask whether OLPC has reversed its open software ideology. It has also resulted in skepticism about whether the chief purpose of the program is to improve the lives of children in developing nations or produce marketable devices for global educational settings (see James 2010).

3.4 Sen, Latour and Liberation

Our purpose in this section is to employ the above empirical sketch of OLPC as a space for deepening the exploration of the conceptual overlaps between Sen and Latour. OLPC is an especially productive case as it highlights the importance of analyzing encounters between developmental designs and their intended users. We will argue that the OLPC program helps to bring out some potential problems in Sen's approach to technology and that these problems may be resolved through supplementing Sen's notion of 'conversion' with Latour's notion of 'translation'.

Let us start, then, by noting that some of the features of OLPC seem to suggest that it is consistent with the 'liberation from' orientation described in the previous sections. For the developers of OLPC, capability and liberation are achievable through increasing the ability of individual children to safely ignore and transcend their immediate circumstances. For instance, the promotional video on the OLPC website stresses the individual ownership of computers, arguing that children must be free to use them whenever and wherever they want. Only after stating this condition does the video briefly explore the social possibilities of computers by discussing how children may use them to 'collaborate on projects'.[8] Likewise,

---

[8] http://laptop.org/en/vision/index.shtml

OLPC tends to equate laptop ownership with education. This is made evident when the voiceover in the promotional video asks "Why give a laptop to a child who may have no electricity or even running water? That's a very good question. But if you substitute the word 'laptop' with 'education', the answer becomes clear..." The laptop is presented as a piece of technology that gradually frees the learning child from a series of material necessities, such as school buildings (the computer is rainproof and its screen works in direct sunlight) and electricity supplies (the computer is charged by hand-cranked power).

Through these designs, the learning child emerges as self-directed and maximally separate from the immediate environment. One problem with this view is that its underlying understanding of 'freedom' easily obscures the differing importance of the multiple relations that constitute learning. Through its focus on the self-teaching and self-sufficient laptop-child, OLPC seems to confuse 'freedom' with the use of laptops and the digital spaces of computer networks, therefore bypassing other important relations that make learning possible. The promotional materials rarely mention teachers, properly functioning school facilities and the social and educational importance of sharing learning equipment (see James 2010).

We will return to these questions shortly, but let us first attend to a possible lacuna that the OLPC case exposes in Sen's work. From our viewpoint, Sen's approach alone does not appear sensitive enough to the socio-technical complexities of the OLPC program. Sen is chiefly concerned to register and encourage the movement of people from subaltern status to self-determining agents. He seems less able to recognize that there are many situated and conflicting forms of liberation and many different ways to configure an agent. Sen does not provide us with tools for exploring how assumptions about human value are folded into designed materials and how they shape and become shaped by actual users. This leaves unexamined the multiple modes of agency at work both through the OLPC program and in the settings where it is implemented.

Understanding these tangled socio-technical agencies is crucial from a critical perspective, as the 'freedom from' orientation of OLPC clearly has 'biopolitical' implications for majority world states (see Bull 2007; Rose 2007). Even as OLPC appears to assist in the state function of education, it also sidelines the state by seeking to transcend it in various ways. Our account of OLPC suggests that the program is underpinned by a rather individualized notion of education, inviting children to manage their own learning while abstracting away some of the essential relations that constitute everyday settings. In the short term, the state might enjoy the benefit of an increasingly literate, numerate and well-informed young population, but in the long term, the concern is that states might not be understood as potential sites of liberation by their own populations. This is problematic because developing capabilities of populations often go hand in hand with the developing capacities of states to govern effectively. However, we are not suggesting that all majority world states are sites of liberation or that having access to global flows of information will necessarily draw the aspirations of young people away from their home countries. We argue that the capabilities approach developed by Sen needs supplementing if it is to clearly distinguish itself from the conflations of individual and collective liberation that appear in the OLPC imagination. Next we will show that this can be done by enhancing Sen's notion of 'conversion' with Latour's notion of 'translation'.

3.4.1 Sen's Moment of 'Conversion'

We have seen that Sen (1999) argues for an account of wellbeing that treats human variation as the norm. He is concerned that a strong focus on utility is insensitive to the real life conditions of developing nations. For Sen, wellbeing needs to be understood as the result of interactions between commodities and the specific circumstances and characteristics of individuals. This argument is best illustrated by Sen's (1983, 1985) familiar example of the bicycle, which shows that the possession of a bicycle has different consequences for an individual who has use of legs and another who does not. Although both persons possess the same commodity, it enables one of them to travel while the other remains immobilized, thereby reminding us that possession of a bicycle does not directly lead to wellbeing.

The bicycle example encapsulates a series of productive concepts that Sen develops throughout his work. After Sen (1999), we may call the act of putting the bicycle into situated use a moment of 'conversion', where the commodity is adapted to personal needs. This conversion, in turn, contributes to a person's 'functionings': the actual 'beings' and 'doings' that improve his/her wellbeing. For instance, in the case of the bicycle, such functionings might include the possibility of a person to travel to work or visit friends, thereby enabling him/her to earn income and sustain social relationships. These functionings compose the 'capabilities' of an individual, reflecting "the person's real opportunities or positive freedom of choice between possible life-styles" (Clark 2005, 1343). Capabilities, then, stand for the actual 'freedom' a person has to decide and achieve his/her goals and plans. Taken together, functionings and capabilities offer a broad and empirically sensitive evaluative space that highlights the specific ways in which human agency and wellbeing is shaped in various settings.[9]

The concept we find most promising, both in terms of empirical analysis and evaluation, is 'conversion', which, as we will indicate in a moment, resonates with Latour's work on 'translation'. Conversion describes those instances when a person 'converts' commodities or technologies into situated uses, thereby adapting them for personal needs. Whatever the purpose of conversion, Sen's account of wellbeing is clearly less concerned to measure individual possession than to chart changes that

---

[9] Sen's reluctance to list crucial capabilities distinguishes him from philosopher Martha Nussbaum, who has engaged in the normative project of defining central capabilities for evaluative purposes. David A. Clark (2005, 1346) argues that this is a strategic move from Sen, who, in allowing the capabilities approach to remain incomplete, sidesteps "the charge of paternalism by leaving each and every person with the freedom to decide his/her own set of functionings."

lead people to gain greater understanding of what they can become. Sen's focus on such productive shifts in human agency is also evident in his detailed attention to the various personal, social and environmental features that impact an individual's possibilities to actually convert commodities into functionings (Sen 1985). These 'conversion factors' might include, among other things, the gender, age, health and social status of a person (see Crocker and Robeyns 2010).

At the same time, we think that two aspects of conversion in particular need to be specified to make the concept compatible with the concerns of technology and design based developmental projects. First, despite discussing various social and environmental features of conversion, Sen (1985, 1999) is rather focused on the human individual performing the conversion, while surrounding elements and people only appear as resources or obstructions to this process. We argue that there remains room to expand the notion of conversion with a more relational view, where designed materials are not converted by individuals or even humans alone, but by situated arrangements of people, technologies and environments. Second, Sen describes conversion mainly as a linear process, where persons adapt commodities to reach certain pre-planned goals. This seems a fairly reductive and instrumental notion of human action and development, both of which tend to have more open-ended and uncertain qualities. We regard Latour's notion of 'translation' as helpful in dealing with these problems, but before exploring this argument, we will briefly return to OLPC in order to exemplify our ideas.

3.4.2 OLPC: Second Empirical Interlude

Emerging research on the implementation of OLPC provides fascinating glimpses into the actual interaction between the XO laptops and their intended users, the children and teachers of developing countries. Let us start by considering a pilot study involving the pupils of a rural Uruguayan elementary school, conducted by Juan Pablo Hourcade, Daiana Beitler, Fernando Cormenzana and Pablo Flores (2008). During fieldwork, the researchers found out that the laptops were well received by teachers and children alike, inspiring pupils to read and write more and explore new computer programs. At the same time, the children rarely accessed web pages or communicated with anyone outside Uruguay. This may be explained by poor connectivity and language barriers. As the researchers note, children were more interested in how the laptops entwined with their local settings—for instance, the pupils 'customized' laptops with stickers, gathered around them in small groups and voiced their discontent about the music making software that lacked features from their own popular music.

Another empirical study was conducted by Hermann Härtel (2008), who observed a series of OLPC pilot projects in rural and urban schools in Ethiopia. According to Härtel, Ethiopian children required constant support from their teachers while working on their laptops, as managing the track pads and opening specific applications was experienced as difficult. Even the educators appeared uncomfortable with the computers, mostly due to lack of proper training, but also because of the unpaid extra work that the implementation of XO laptops demanded from the already stressed teachers. For this reason, Härtel recommends a stepwise approach in future implementations of OLPC: internet and application use should be postponed until children and teachers have enough familiarity with computers and until there are sufficient funds and infrastructures to support such a move. For the moment, the most important contribution of the XO computers is that they enable children to access electronic reading materials, because books are scarce in Ethiopian schools.

David Hollow (2009) has also studied experiences with XO laptops in Ethiopia. He stresses that the greatest challenge for OLPC is to incorporate the XO laptops into local curricula. Hollow claims that the OLPC program should pay more attention to teacher involvement and seek to minimize the radical shifts in local pedagogical practices that its constructionist ideology suggests. The laptops received positive response from Ethiopian schools, but some teachers felt that the computers were a distraction from classroom activities, as the pupils often used laptops for playing and chatting rather than for educational purposes. This observation was confirmed by the children, who admitted that they preferred using the laptops for non-curricular activities. Children were also worried about the laptops: they feared that they might get stolen or broken by others, which made some leave their laptops at home.

3.4.3 Latour's Moment of 'Translation'

The above encounters between XO laptops and educational settings remind us that moments of conversion often appear more 'entangled' than the writings of Sen suggest. It seems difficult to distinguish between personal, social and environmental conversion factors, as the implementation of laptops mobilizes a complex arrangement of teachers, children, classrooms, funding and infrastructures. At the same time, it is clear that individuals require help from others to convert commodities and technologies into functionings. For instance, the children in Ethiopia needed teachers to support their use of the laptops, and the teachers needed training, financial support and technological infrastructure to be able to help the children. For this reason, we think that it is more productive to approach conversion as a collective rather than as an individual accomplishment.

The above accounts also suggest that conversion is an ongoing and uncertain process. Encounters between children and laptops produced a range of doings that were not necessarily pre-planned, either by the developers or by the children or teachers. For instance, pupils in Uruguay employed the laptops to sustain local relationships rather than to simply make 'global' connections, as the OLPC promotional materials often stress. Likewise, some Ethiopian children left their laptops at home to protect them from others. Such conversions created multiple uses for XO laptops, some reinforcing, others contradicting one another. The XO laptops may offer children moments of enjoyment during the long school day by allowing them to play games and chat with peers. At the same time, however, such 'fun' uses may frustrate the work of teachers by diverting children's attention from curricular activities. As David A. Clark (2005) points out, Sen has not explored the possibility of these 'negative functionings' in his work. Nevertheless, studies on OLPC in Uruguay and Ethiopia clearly indicate that conversions may result in a meshwork of interacting functionings that are 'good' for some persons and 'bad' for others.

On our view, Latour's (1988, 1999, 2005) concept of 'translation' is a productive tool for bringing out such collective and unpredictable aspects of conversion. According to Latour (1999, 311), translation points to "the work through which actors modify, displace, and translate their various and contradictory interests." The term originates in the writings of Michel Callon (1980, 1986), who adapted it from philosopher of science and technology Michel Serres (1974). Seeing translation as a basis for culture, Serres employs the notion to highlight the constant work involved in establishing relations between disparate elements and domains, whether these are texts, cultures, or disciplines. For Serres, translation always transforms the elements it brings into relation with one another, and it is therefore necessarily an innovative event that gives rise to new meanings, associations and agencies.

Callon mobilized the concept of translation to illuminate the tangled relations between science, technology and society in case studies on the development of an electric car (Callon 1980) and a marine biological conservation project (Callon 1986). Both studies indicate that technological practices depend on the careful alignment of diverse elements, from artefacts and infrastructures to engineers and legislators. 'Translation' is the process whereby such heterogeneous components are brought together in an attempt to transform their varied interests and properties into a common orientation. This process is never straightforward, as the humans, ideas and materials involved constantly shape one another, not only generating complications but also new forms of association and agency.

Latour has since developed this idea in his own work, indicating how organizing and sustaining any scientific or technological project—from urban transportation systems to the circulation of scientific knowledge—requires translation work: an uncertain and ongoing composition of a 'collective' of humans and non-humans that need to be studied and understood in its empirical specificity (see Latour 1988, 1999, 2004, 2005).[10] We think that the notion of translation is especially useful when engaging with the complexities of design interventions like OLPC. David

---

[10] Latour has often elaborated this idea in the shape of the notion of 'mediators' which he separates from 'intermediaries' (Latour 1999, 2005). Intermediaries reflect the way technologies and artefacts are often conceptualised in social sciences: as instrumental means for purposeful human action. On the contrary, the notion of mediator suggests that any relation between, for instance, a user and a technological innovation is specific and should be studied accordingly. Mediators "transform, translate, distort, and modify the meaning or the elements they are supposed to carry" (Latour 2005, 39). To see something as a mediator directs attention to the specific ways in which human and non-human materials form relations "that did not exist before and that to some degree modifies the original two" (Latour 1999, 179).

Mosse (2005, 9) has argued that translation can be seen as describing whole developmental projects, where "heterogeneous entities—people, ideas, interests, events and objects (seeds, engineered structures, pumps, vehicles, computers, fax machines or databases)—are tied together by translation of one kind or another into the material and conceptual order of a successful project". This argument suggests that we can only talk about a 'succesful' project as a situated outcome of ongoing and mutual transformations among its diverse actors and elements. The notion of translation therefore encourages us to think about development as a continuous 'cultivation' of relations. We are, in other words, invited to recognize and accommodate the multiple and unexpected forms of togetherness and agency that emerge in actual encounters between the persons, interests and materials of developmental projects.

Seen in this way, it becomes important not only to foster freedoms but also to sustain socio-technical *collectives* that provide space for such freedoms. This is exactly what most of the above studies on OLPC suggest: the children and their communities were elaborating highly complex relational 'functionings' by using laptops for negotiating ties to their immediate environments. Such practices of shared technology use indicate that converting designed materials into functionings is not entirely an individual or goal-oriented act. Rather, functionings are made possible in unpredictable entanglements of people, artefacts and environments, where 'liberation within'—engaging in embedded processes of learning together with others—is often more important than reaching an abstracted state of 'liberation from'.

At the same time, the notion of translation helps us to recognize the 'geographies of responsibility' (Massey 2004) that crisscross developmental interventions and tie together people, interests and materials across spatial and temporal scales. We have seen that OLPC may be criticized for its attempt to provide a quick technological fix to educational problems in developing nations while paying less attention to the inevitable need for proper training, infrastructure and financial support to sustain the everyday use of XO laptops (e.g. Hollow 2009; Härtel 2008; James 2010). If we view the implementation of OLPC as requiring ongoing translation work, we become more sensitive to the fact that developmental projects entail long-term commitments across local–global divides, from designers to governments and all the way to specific communities. Highlighting translations, then, enables us to broaden the developmental space for individual and social 'growth', both of which tend to deviate from rather than conform to clear-cut plans of progress (see also Lee 2005).

3.5 Conclusion

We have employed the One Laptop per Child (OLPC) program as an empirical arena for staging an encounter between Amartya Sen and Bruno Latour, offering insight into some of the differences and commonalities between the two authors. It is clear that Sen's focus on 'freedom' and 'development' is considerably enriched by Latour's critical attention to the actual shaping of human agency in various socio-technical entanglements. Likewise, Latour's 'distributed' notion of the 'human' may be complemented by Sen's sensitivity to questions of global justice and human rights, inviting us to think about more responsible ways of carrying out developmental design projects.

We have demonstrated these productive outcomes of the Sen/Latour encounter by exploring parallels between Sen's concept of 'conversion' and Latour's concept of 'translation'. We have shown that both authors share an interest in the interactions between persons and designed materials as well as in how such interactions 'convert' and 'translate' both people and materials to better adapt to specific settings. We have also argued that Sen's tendency to emphasize individual agency at the expense of collective action may be supplemented by Latour's strongly relational notion of agency. Latour stresses that the more attachments we form with our surroundings, the more possibilities we have to act and realize our goals. This view enables us to bring out the role of design in the composition of developmental collectives, as a collective is never a 'purely' human accomplishment but mediated by various non-human materials (see also Oosterlaken 2009, 2011).

Taken together, Sen and Latour invite a systematic empirical analysis of developmental projects in terms of 'liberation within' rather than 'liberation from'. This means that the *quality* of the relations among persons, technologies and environments is more important than minimizing dependencies between people and their surroundings. Above all, such a relational view opens new possibilities to develop more sustainable and lasting connections between persons, technologies and the 'world'—a project we have affirmed by strengthening connections between design and development, Bruno Latour and Amartya Sen.

References

Bull, M. (2007). Vectors of the biopolitical. *New Left Review, 45*, 7–25.
Callon, M. (1980). Struggles and negotiations to define what is problematic and what is not: The sociology of translation. In K. Knorr, R. Krohn, & R. Whitley (Eds.), *The social process of scientific investigation: Sociology of the sciences yearbook*. Dordrecht/Boston: Reidel.
Callon, M. (1986). Some elements of a sociology of translation: Domestication of the scallops and the fishermen of St Brieuc Bay. In J. Law (Ed.), *Power, action and belief: A new sociology of knowledge? Sociological review monograph*. London: Routledge/Kegan Paul.
Clark, D. A. (2005). Sen's capability approach and the many spaces of human well-being. *Journal of Developmental Studies, 41*(8), 1339–1368.
Crocker, D. A., & Robeyns, I. (2010). Capability and agency. In C. W. Morris (Ed.), *Amartya Sen* (pp. 60–90). Cambridge: Cambridge University Press.
Härtel, H. (2008). *Low-cost devices in educational systems: The use of the "XO-laptop" in the Ethiopian Educational System*. Report distributed by the Division of Health, Education and Social Protection, Information and Communication Technologies, GTZ-Project, Deutsche Gesellschaft fur Technische Zusammenarbeit, Eschborn, Germany. Retrieved from www.gtz.de/de/dokumente/gtz/2008-en-laptop.pdf
Hollow, D. (2009). *Initial reflections on the Ethiopia XO 5000 Programme* (Working Papers). Egham: Royal Holloway University of London. Retrieved from www.gg.rhul.ac.uk/ict4d/workingpapers/HollowXO5000.pdf

Hourcade, J. P., Beitler, D., Cormenzana, F., & Flores, P. (2008). Early OLPC Experiences in a Rural Uruguayan School. In *Extended Abstracts of CHI 2008 Conference* (pp. 2503–2512). New York: ACM Press.

James, J. (2010). New technology in developing countries: A critique of the One-Laptop-per-Child Program. *Social Science Computer Review, 28*(3), 381–390.

Kraemer, K. L., Dedrick, J. & Sharma, P. (2009). One Laptop per Child: Vision vs. Reality. *Communications of ACM, 52*(6), 66–73.

Latour, B. (1988). *The pasteurization of France.* Cambridge: Harvard University Press.

Latour, B. (1993). *We have never been modern.* Cambridge: Harvard University Press.

Latour, B. (1996). *Aramis or the love of technology.* Cambridge: Harvard University Press.

Latour, B. (1999). *Pandora's hope. Essays on the reality of science studies.* Cambridge: Harvard University Press.

Latour, B. (2004). *Politics of nature: How to bring the sciences into democracy.* Cambridge: Harvard University Press.

Latour, B. (2005). *Reassembling the social. An introduction to actor network-theory.* Oxford/New York: Oxford University Press.

Latour, B., & Hermant, E. (1998). *Paris ville invisible.* Paris: Les Empecheurs de penser en rond: La Découverte.

Law, J., & Hassard, J. (Eds.). (1999). *Actor network theory and after.* Oxford: Blackwell Publishing.

Lee, N. (2005). *Childhood and human value. Development, separation and separability.* Maidenhead: Open University Press.

Massey, D. (2004). Geographies of responsibility. *Geografiska Annaler: Series B, Human Geography, 86*, 5–18.

Morris, C. W. (Ed.). (2010). *Amartya Sen.* Cambridge: Cambridge University Press.

Mosse, D. (2005). *Cultivating development. An ethnography of aid policy and practice.* London: Pluto Press.

Negroponte, N. (1995). *Being digital.* New York: Alfred A Knopf.

Oosterlaken, I. (2009). Design for development: A capability approach. *Design Issues, 25*, 91–102.

Oosterlaken, I. (2011). Inserting technology in the relational ontology of Sen's capability approach. *Journal of Human Development and Capabilitites, 12*(3), 425–432.

Papert, S. (1980). *Mindstorms. Children, computers, and powerful ideas.* New York: Basic Books.

Rose, N. (2007). *The politics of life itself. Biomedicine, power, and subjectivity in the twenty-first century.* Princeton: Princeton University Press.

Sen, A. (1983). Poor, relatively speaking. *Oxford Economic Papers, 35*, 153–169.

Sen, A. (1985). *Commodities and capabilities.* Oxford: Oxford University Press.

Sen, A. (1999). *Development as freedom.* Oxford: Oxford University Press.

Sen, A. (2010). *The idea of justice.* London: Penguin.

Serres, M. (1974). *Hermès III. La traduction.* Paris: Editions de Minuit.

Chapter 4
Evaluating Emerging ICTs: A Critical Capability Approach of Technology

Yingqin Zheng and Bernd Carsten Stahl

4.1 Introduction

The present chapter sets out to provide a theoretical perspective for evaluating social implications of technology and to give some examples of how to apply it. Drawing on two distinct but related bodies of theory, namely the capability approach and critical social theory in information systems, which share some commonalities and also complement each other (Zheng and Stahl 2011), the chapter proposes the Critical Capability Approach of Technology (CCAT) as a novel way of evaluating technologies. This approach to evaluation will be useful for policy makers, technology designers and developers and consumers who have to consider the social consequences of technologies.

The CCAT is applied in this chapter to discuss the implications of emerging information and communication technologies (ICTs), which are difficult to define and assess prior to their adoption. In most cases the development and use of emerging ICTs do not follow simple and linear patterns. Uses and consequences are often ambiguous and contradictory. ICTs can be seen as "Janus-faced" because of these ambiguities (Arnold 2003). Despite this unclear nature of emerging ICTs – or maybe even because of it – and because of the broad use and wide reach of such technologies, it would be important to evaluate such new technologies from a theoretically sound perspective in order to address possible problems as early as possible.

Y. Zheng (✉)
School of Management, Royal Holloway, University of London, Surrey, UK
e-mail: yingqin.zheng@rhul.ac.uk

B.C. Stahl
Faculty of Technology, Centre for Computing and Social Responsibility,
De Montfort University, Leicester, UK

The rest of the chapter starts by outlining the Critical Capability Approach of Technologies consisting of four principles of evaluating social implications of technology. It will then introduce emerging ICTs and describe three examples in some more depth: affective computing, ambient intelligence, and neuroelectronics. Possible implications of these examples of emerging ICTs are then explored from a CCAT perspective. The chapter concludes by synthesizing how our approach goes beyond current work and pointing to future direction of research.

4.2 The Critical Capability Approach of Technology

This section briefly introduces our approach, which is based on Sen's Capability Approach and the authors' research on critical theory of ICT. Sen's capability approach is essentially concerned with substantive freedoms, i.e. the real opportunities people have to lead the life they have reason to value (Sen 1999). Despite its huge contribution to the field of development economics and ethics, the capability approach was not often used to evaluate the design and consequences of technologies until recent years. Zheng (2009) draws upon the capability approach to question the space in which technology should be evaluated in relation to human development and proposes a capability approach perspective of e-development. Similarly, Oosterlaken (2009) proposes to incorporate the philosophical insights of the capability approach into design science towards a "capability-sensitive design" approach.

It is necessary to note that capability approach has been deliberately kept vague and "incomplete" (Robeyns 2006), and requires further theorization for researchers who seek to apply it in specific areas. One of the challenges in applying the capability approach to study technology is that there is no explicit theorization of technology by Sen or other theorists active on the topic, given the root of the capability approach in economics and development ethics. There is thus a risk that the capability approach is adopted with a simplistic perspective on technology by regarding it as a type of commodities, i.e. goods and resources that can be readily drawn upon by users (Zheng 2009; c.f. Oosterlaken 2009).

In response, Zheng and Stahl (2011) compare Sen's capability approach and the studies that apply Critical Theory (CT) to technology or information systems. They identify three main areas in which the capability approach can be enriched by learning from the CT: the conception of technology, the conception of agency and its "situatedness", and the methodological implications of these two conceptions. The comparison is based on the premise that the capability approach and CT share a significant array of commonalities: they both constitute schools of thought that are meant to make a difference – to improve individual and social lives; both are normative theories rooted in ethics, and develop different streams of ideas to support freedom, empowerment and emancipation. In this chapter, we build upon these insights to propose an integrated Critical Capability Approach of Technology, which aims to provide a set of principles that can be used to evaluate the design and impact of technologies.

Table 4.1 Comparing key concepts of the capability approach with critical theory

	Sen's capability approach	Critical theory
The ends of technology	Expansion of freedom, or removal of unfreedoms that restrict individuals from exercising their reasoned agency	Emancipation, or removal of injustice, alienation and domination
Human diversity	Attention to diversity of and discrepancies in human conditions; questions what conversion factors are in place to generate potentials to achieve, and to allow people the freedom of choice to realize the achievement	Special attention is given to those individuals who are oppressed or alienated. These are usually those who are least able to defend themselves and lack political and social power and representation
Individual agency	Central to the capability approach, together with the concept of well-being, agency forms the basis of addressing deprivation; embedded in socio-cultural conditions	Emphasis on the effect of social structures on individual agency, especially through hegemony of ideology
Technology	Regarded as commodities, i.e. goods and resources, meaningful only in terms of their contribution to people's capabilities; means rather than ends	CTICT highlights ideological qualities and hegemonic functions of technology; sensitive to interpretive flexibility of technology and its role in distribution of power
Methodology	Emphasis on bottom-up, participatory approach	Sensitive to power and political issues; emphasis on participation, and reflexivity; sensitive to reification and hegemonic potential of knowledge and methodologies

4.2.1 Comparing Sen's Capability Approach with Critical Theory of ICT

Table 4.1 presents and compares the key concepts from Sen's capability approach and critical theory, which draw upon previous work on the topic (Zheng 2009; Zheng and Stahl 2011). We will briefly elaborate each of the concept from both perspectives.

4.2.1.1 The Ends of Technology

The capability approach is ultimately concerned with the actual experience of individual lives, where the whole approach is grounded and points to – "the expansion of freedom ... both as the primary end and as the principle means of development" (1999, p. xii). The aim of the capability approach is to enhance people's opportunities to lead a life as they have reason to value. From this perspective, technology is seen as a means to expand human freedom, and thus should be evaluated in terms of its contribution to an individual's capability set in terms of both well-being freedom and agency freedom (Sen 1992).

Critical theory similar seeks to build a better world. What is wrong about the world from the critical perspective is that human beings are not given the opportunity to live the best possible lives they could or to achieve their potential (Robeyns 2005, p. 163). The term "emancipation" denotes the attempt to give people this opportunity, to allow them to live up to their potential. It is difficult to clearly describe what constitutes emancipation. Hirschheim et al. (1995, p. 83) define emancipation as "all conscious attempts of human reason to free us from pseudo-natural constraints." It is plausible to assume that alienating circumstances can be found in the environment as well as the agent. Emancipation therefore needs to address issues of "false or unwarranted beliefs, assumptions, and constraints" (Ngwenyama and Lee 1997, p. 151).

A critical theory of ICT (CTICT) (Zheng and Stahl 2011) thus stands for the recognition that ICT has the potential to improve social reality and promote emancipation, but often has opposite effects. Critical research aims to address this by epistemological means (e.g. by exploring the nature and consequences of ICT), but aims to go beyond this. Its awareness of the socially constructed nature of technology and its ability to describe interpretive flexibility[1] together lead to a sensitivity of the relationship between technology and power. Social and economic structures influence the values on which technologies are built and thereby the affordances they offer to users. These insights have both theoretical and practical relevance and would be able to inform our development and use of technology in order to contribute to the development of a better society.

4.2.1.2 Human Diversity

Essential to the capability approach is the recognition of human diversity which gives rise to an explicit differentiation between "spaces of equality". As Sen observes:

> We are deeply diverse in our internal characteristics such as age, gender, general abilities, particular talents, proneness to illness, and so on) as well as in external circumstances (such as ownership of assets, social backgrounds, environmental predicaments, and so on). (Sen 1992, p. xi)

Equality in one space to lead a valuable life, e.g. income, does not necessarily mean equality in life opportunities to achieve it, e.g. access to quality healthcare.

---

[1] Interpretive flexibility stands for the view that technology is not objectively given but is socially constituted through perception and use. This means that a particular technical artefact may have different means and uses in different contexts (Doherty et al. 2006). There are different views on whether this refers to the nature of technology per se or only to its perceptions (Cadili and Whitley 2005). For this purpose these distinctions are not of central importance because both positions are compatible with a critical perspective.

Therefore, individual variations, including structural differences in society, must be considered when developing technologies which aim to improve individual opportunities to take a full part in society.

Furthermore, Sen emphasizes the great variability in the instrumental relation between low income and low capability, affected by individual, social and environmental diversities. For example, the elderly or the disabled groups will encounter greater "conversion" difficulties in converting income into functionings. A relatively poor person living in a rich country may be more deprived in terms of capabilities compared to people with lower absolute income but living in less opulent countries. By the same token, there is no absolute linkage between availability of technology and level of capability. Advanced technologies do not necessarily improve an individual's quality and chances of life, because they require more complex conversion factors to be exploited. The digital divide does not lie between those have or have not access to technologies, but between those who are in a position to convert relevant and appropriate technologies into capabilities and those who are not.

The critical theory of ICT does not seem to touch upon the diversities of human conditions, but treats emancipation as a universal pursuit.

4.2.1.3 Situated Agency

Sen (1985) defines agency as the freedom to set and pursue one's own goals and interests, which may also include furthering the well-being of others, respecting social and moral norms, or acting upon personal commitments and the pursuit of a variety of values. A person is thus viewed as an "agent", as opposed to a "patient" whose well-being or the absence of well-being is the only concern (Robeyns 2005). The recognition of individual agency does not imply an unconditional acceptance of whatever a person happens to perceive as valuable, or acceptance with as much intensity as valued by the person. Scholars of the capability approach have pointed out the issue of "adaptive preferences" and "restricted agency" (Peter 2003). Even then there is still space, Sen also argues, for the agency to be evaluated and appraised (Sen 1985). Yet Sen does not go on further to discuss what the possible constraints on human agency are and what criteria should be used to appraise individual preferences, apart from emphasizing the role of public discussions.

While critical scholars don't tend to deny the possibility of individual agency, they point to the structural conditions of individual agency. Of primary importance among those conditions is the economic constitution of society, or, to put it differently, the way in which capitalist systems structure agents' options. Critical scholars tend to point to the importance of historical backgrounds in understanding social situations. They tend to underline the importance of social structures in enabling or denying emancipation, that is, the ideological character of social structures which limit personal freedoms. In other words, critical theory seeks to reform institutions in order to remove any injustice and to reach a better world.

Deneulin (2006) draws upon Ricoeur's ethical vision to improve Sen's notion of individual agency to that of *social-historical agency*, which refers to what human beings can really do or be given the particular socio-historical structures in which they are living. Zheng and Stahl (2011), drawing upon critical theory of ICT, propose the concept of *situated agency* to express the idea that individual agency is not only a product of specific socio-historical settings, but also situated in a sometimes invisible or taken-for-granted network of ideology. Critical theory also teaches us that individuals participate in the production and reproduction of these socio-historical structures and ideological tenets.

Such a conception of restricted agency has important implications for ICT and social development, as it gives rise to a sensitivity towards deep-seated power structures and rationalities. For example, a participatory approach to development, which Sen himself strongly advocates and which is popular in most development projects, may disguise or even strengthen incipient articulation of power embedded in social and cultural practices, hence the "tyranny of participation" (Cooke and Kothari 2001). It is possible that participatory methodologies may reify existing inequalities and affirm the agenda of elites and other more powerful actors (Kothari 2001). A critical capability approach that conceptually and methodologically incorporates *situated agency* as a key element would allow us to critically evaluate social arrangements and cultural norms as part of the assessment of technologies' role in enhancing individuals' well-being and agency freedom.

4.2.1.4 Conception of Technology

Sen's capability approach does not touch upon the role of technology, which is not surprising. Work that incorporates technology into Sen's capability approach often makes an implicit assumption of technologies as goods and resources that are independent of values and beliefs. For example, Zheng (2009) proposes a view of seeing ICT as commodities. Such a view implies that technology is neutral and can be readily drawn upon to serve the purposes of human development.

In contrast, critical theory has a long history of being applied to the study of technology and information systems (Stahl 2008a). Critical theory of ICT (CTICT) emphasizes the importance of revealing technology's ideological qualities and hegemonic functions. Ideologies may be socially accepted views which are part of all collective constructions of reality and therefore a necessary consequence of a social constructivist worldview. An example of ideology is pursuit of modern life in societies. It is possible that ideologies may have positive consequences when they allow for the development of positive views of experiences (McAulay et al. 2002).

Technology can serve as hegemonic means by supporting and rendering invisible such ideologies (Saravanamuthu 2002; Feenberg 1999). At the same time, technology itself can have an ideological status, for example when technology is equated with progress and progress is assumed to be unquestionably desirable; when technology represents "expert knowledge" that exercises "disciplinary power" (Foucault 1980); or when technology embodies contested social regulations, for example through digital rights management. The ideological quality of technology can be upheld by hegemonic means which remove technology from questioning. Such hegemonic means may be customs, social consensus or the law.

Work in CTICT is also concerned with the role technology plays in areas and issues of alienation and oppression, and with issues that affect emancipation or its potential. Power-related issues are of prime interest, in particular those where power is related to technology. Control and surveillance technologies are good examples.

4.2.1.5 Methodological Issues

Capability approach-based methods seem to be uniquely suited to describing technologies and evaluating different options. Many ICTs have potentially far-ranging consequences and design decisions that aim to be conducive to justice and emancipation have to rely on some sort of measure that will allow comparisons of different options or outcomes. Examples of this could include design decisions in the development of technologies such as ambient intelligence applications or action choices in technically relevant social action. Such work could be supported and underpinned by critical perspectives, for example by questioning participants' or experts' opinions or by conducting ideology critiques of capability measures.

The methodologies typically employed within critical theory can also be helpful in dealing with ICT. Critical work that looks at the linguistic construction of technology is important to unpack black boxes that determine affordances and mediations of technology. This type of work is closely aligned with some of the work currently done in ethics and ICT, such as disclosive ethics (Brey 2000; Introna 2005). An understanding of how language is used to portray particular technologies and projected developments can also be conducive to better design of technologies, such as suggested by the idea of value-sensitive design (van den Hoven 2008).

4.2.2 The Four Principles

The above comparison between Sen's capability approach and critical theory provides a basis for the following four principles that constitute the Critical Capability Approach of Technology:

1. The principle of human-centered technological development;
2. The principle of human diversity;
3. The principle of protecting human agency;
4. The principle of democratic discourse;

4.2.2.1 The Principle of Human-Centered Technological Development

The fundamental principle of technological development is to enhance substantive freedom, i.e. people's capabilities to lead a life they have reason to value, and to remove unfreedom and injustice. Neither the capability approach nor critical theory specifies, or believes that it is possible or desirable to specify, an ideal society. Both, therefore, focus on generating opportunities and removing barriers for emancipation, freedom and justice in specific contexts. This fundamental principle leads to the idea that technology is means to an end but never ends in itself, hence not intrinsically desirable or beneficial.

This fundamental principle of technological development is reflected in the discussion of "human-centered design" by Buchanan (2010), who observes that

> [w]e tend to discuss the principles of form and composition, the principles of aesthetics, the principles of usability, the principles of market economics and business operations, or the mechanical and technological principles that underpin products

and often neglect that "design is grounded in human dignity and human rights" (Oosterlaken 2009). Indeed, most modern day technologies are driven by market forces, with the aim to fulfill people's desire, comfort, or "happiness". In contrast, the capability approach is essentially concerned with both well-being and agency freedom, or effective possibilities of realizing achievements and fulfilling expectations. Sen (1985) emphatically distinguishes "valuing" from "being happy with" or "desiring" – "valuation is a reflective activity in a way that 'being happy' or 'desiring' need not be" (p. 29–30).

Similarly, critical theory questions individuals' "false consciousness" and does not consider desire as a rational basis to build society, as individuals often desire things and have preferences that are arguably not in their own best interest. Simple examples of this are the abuse of alcohol and drugs. The problem goes beyond this, however, in that individuals can often become complicit in the social structures that oppress and alienate them. This phenomenon is often described under the heading of "hegemony" (Stahl 2008b). It raises the question which criteria are to be applied to determine whether individual preferences are acceptable, and calls for a critical discourse on users' preferences of technologies and the contribution of these technologies to capabilities or emancipation.

4.2.2.2 The Principle of Human Diversity

The second principle of the CCAT is that of human diversity, ranging from individual characteristics, environmental conditions to social arrangements (Robeyns 2005). It is on this basis that Sen asks the important question of "equality of what" (ibid.), and proposes a different "evaluative space" (Sen 1993) that consists of the plurality of functionings and capabilities, as opposed to income, utility or desire-fulfillment in traditional economic approaches. Zheng (2009) argues that "the emphasis of human diversity of the capability approach offers a critique on the unquestioning pursuit of ICT diffusion across contexts, and a tendency to apply universal criteria on using ICTs as developmental instruments".

The principle of human diversity also raises the question on conversion factors that allow people to generate capabilities from particular technologies, as people are endowed with various physical and mental characteristics, live in diverse environments under different socio-cultural conditions – all of these factors will affect what real opportunities a person can realistically enjoy from the adoption of a particular piece of technology. Therefore, the principle of human diversity will render invalid any assumption or claim about universal benefit of technology. Rather, the questions that need to be asked when evaluating a type of technology are: What capabilities does the technology contribute? For whom? Under what circumstances? What are the enabling factors and what are the barriers?

4.2.2.3 The Principle of Protecting Human Agency

Sen uses the term "agent" to refer to "someone who acts and brings about change, and whose achievements can be judged in terms of her own values and objectives, whether or not we assess them in terms of some external criteria as well" (Sen 1999). To stress the importance of the agency aspect of the capability approach, Crocker (2008) proposes the label of the "the agency-focused capability approach" or "an agent-oriented approach". Even though Crocker restrains from assigning moral priority to agency over well-being, both being essential to one's capability set, he does attach great importance to moral freedom, which implies "a prima facie duty to promote our own agency and that of others in relation to inner compulsions and autonomy-eroding behavior." In other words, improvement of well-being does not automatically justify the deprivation of autonomy of individuals, and people should always retain the power as an autonomous being to be one's own master, to make decision on one's choice in life, whether it relates to the adoption of technology or other choices, such as the political arrangement of a society.

Enhancing human agency is conceptually closely linked to the core category of "emancipation" that is at the core of critical theory. Emancipation in critical research stands for the elimination of "causes of unwarranted alienation and domination and thereby enhance the opportunities for realizing human potential" (Klein and Myers 1999, p. 69). An important implication of the principle of protecting human agency is the resistance to the reification of technology. Reification stands for the process of rendering a socially malleable phenomenon into an apparently "objective" thing (*res*=thing, Latin). Reification of technology refers to perceiving it as merely material artefacts with inscribed, unequivocal characteristics, independent from social practices and processes. This is cozily related to technological determinism, the idea that technology has a definite existence and predictable consequences, and often as the determining factor of social and historical phenomena. For example, technologies have often been reified as productivity tools symbolizing efficiency and progress, while in practice the effects of technologies are entangled in social processes and entail emergent and unintended consequences (Orlikowski 2007). For the present chapter it is of importance because reification can limit agency, freedom and emancipation by obfuscating choices that users could have. An example of such reification of technology is the pervasive and unreflective use of surveillance technologies in numerous aspects of social life, such as CCTV, RFID, and various e-business tools, under the name of individual and national security, productivity control, or customer service. Such technologies are often not as effective as expected in enhancing security and business interests, while unintended consequences such as erosion of individual privacy and autonomy, and even discrimination, are often unquestioned or accepted as normal. As Hirchheim et al. (1995, p. 83) put it, reification "suppresses (i.e. through social 'forgetting' or ideology) the human authorship of certain conditions or practices which then appear indistinguishable from natural law. Emancipation proceeds by revealing the sources and causes of the distorting influences which hide alternative ways of life from us."

4.2.2.4 The Principle of Democratic Discourse

One central position shared by several approaches to technology is that it would be desirable to have democratic control over them. Technology is a core determinant of modern societies but, unlike most other aspects of modern life, they are largely removed from democratic control. In light of the potentially significant impact of such technologies, one can ask whether it is appropriate to leave the development of these technologies to market forces. This position is promoted by philosophers of technology (Brey 2008) but also elsewhere, for example in the technology assessment community (Genus and Coles 2005) and even on high policy levels, for example in the European Commission (2006).

Both critical theory and the capability approach consider democratic discourses as pivotal to any social changes. Implementing such democratic control of technology raises numerous problems. One of these is the question which democratic structures would be required to allow exerting influence on novel technologies. This will lead to issues of property and distribution, where technology owners may be averse to sharing influence. Even more fundamentally is the question how to evaluate technology and how to structure democratic discussions about it. How can we know the possible consequences of technology and how do we know what to do in order to address them? There are no easy answers to these questions. It seems reasonable to assume that democratic control of technology will require a continuous process of monitoring and debating developments. It will also require different ownership models of innovation, which need to be opened up to stakeholders. And, importantly, it will require criteria for evaluating technologies, which is what the present chapter seeks to provide.

4.3 Emerging ICTs

Having presented the four principles of evaluating technologies, this section introduces some emerging ICTs which will be discussed later from a CCAT perspective. The material in this section draws on the findings of a research project which undertook an investigation of emerging ICTs with a view to identifying ethical issues these will raise and providing governance recommendations on a European level.[2] The exploration of emerging ICTs can be understood as a piece of foresight research, which means that we are aware of the limitations of our knowledge of the future and we do not claim to know what will be happening. It can thus be seen as an attempt to address the so-called Collingridge dilemma (Collingridge 1981), which stands for the problem that little is known about the social consequences of a technology early in its life cycle when it is easy to change, and once more is known about social consequences the technology tends to be entrenched and difficult to change. The dilemma highlights both the difficulty and the importance of assessing the direction of technological development before it has taken place.

From an initial survey of over 100 technologies, 70 applications and 40 artefacts, the range of emerging technologies were narrowed down to those supported by high-level socio-technical visions that have the potential to change the way humans interact with the world. After rounds of synthesis, review, and revision, a list of technologies were identified as below. The list does not represent all future and emerging ICTs, nor is it complete, comprehensive or entirely internally consistent. Acknowledging these possible limitations, in this chapter we are primarily interested in demonstrating the relevance or our theoretical approach to evaluation of such emerging ICTs.

- Affective Computing
- Ambient Intelligence
- Artificial Intelligence
- Bioelectronics
- Cloud Computing
- Future Internet
- Human-machine symbiosis
- Neuroelectronics
- Quantum Computing
- Robotics
- Virtual/Augmented Reality

Three technologies from this list are selected as examples and explored in more detail: affective computing, ambient intelligence and neuroelectronics.

4.3.1 *Affective Computing*

A broad definition of affective computing is offered by the MIT research group: "Affective Computing is computing that relates to, arises from, or deliberately influences emotion or other affective phenomena" (MIT Media Lab). Affective computing claims to change the way humans interact with the world through changes in human-

---

[2] The ETICA project (http://www.etica-project.eu/) is funded by the European Community's Seventh Framework Programme (FP7/2007-2013) under grant agreement n° 230318.

computer interaction and changes to technically mediated interaction between humans. It is trying to assign computers the human-like capabilities of observation, interpretation and generation of affect features. It is an important topic for the harmonious human-computer interaction, by increasing the quality of human-computer communication and improving the intelligence of the computer (Tao and Tan 2005). Defining features of affective computing are:

Perceiving emotions/affects: (i.e. sensing of physiological correlates of affect). Humans send emotional cues in interactions. Some may be with the agent's control, such as tone of voice, posture or facial expression, and some outside the agent's control as in the case of facial color (blushing), heart rate or breathing. While humans are normally aware of such emotional cues and react easily to them, most current technology does not react to such emotional cues even though they are part of the illocutionary and perlocutionary function of a speech act.

Expressive behavior by computers/artificial agents: refers to the expression of computers in ways that can be interpreted as displaying emotion. Affective expressions can render interaction more natural and thereby easier for users. It can be employed for persuasive purposes as it increases the likelihood of acceptance of computer-generated content.

Emotional cognition: (agent's empathy, understanding of emotional states). This is the next logical step after recognition of emotions. It includes absorbing emotion-related information, modeling users' emotional state, applying and maintaining a user affect model and integrating this model with the emotion recognition and expression. These activities offer the potential of personalizing individual affect models and catering to individual preferences.

4.3.2 Ambient Intelligence

The general idea of ambient intelligence (AmI) is that electronic devices in our homes, offices, hospitals, cars and public spaces will be embedded, interconnected, adaptive, personalized, anticipatory and context-aware. These six features are now described in more detail.

Embedded: It is important to note that AmI is not the outcome of any single technology or application. It is rather an emergent property of several interconnected computational devices, sensors and ICT systems distributed and embedded in the surroundings. The technology disappears into the background and is usually not consciously experienced.

Interconnected: The devices, sensors and ICT systems are (wirelessly) interconnected as well, thereby forming a ubiquitous system of large-scale distributed networks of interconnected computing devices. For example, the miniaturized biosensor systems that monitor vital body variables are connected to an emergency unit which is connected to the ambulance.

4 Evaluating Emerging ICTs: A Critical Capability Approach of Technology

Adaptive: Because there is no stable connectivity to services and information sources in ad-hoc networks, AmI systems can never base their operation on the availability of complete and up-to-date information and services. As a result AmI systems have to organize their services in an adaptive way, i.e. the degree of service varies with the amount of information available and the reach-ability of external services.

Personalized: AmI is personalized to specific user needs and preferences. For example, an interactive interface in one's mirror can provide its user with personalized information about the weather, traffic jams and appointments. In other words, AmI is user-centered.

Anticipatory: AmI can anticipate the desires of its user(s). Consider an AmI system in the context of one's home that monitors behavioral patterns, infers one's mood from the behavioral patterns and adjusts the light and music accordingly. The system is pro-active and anticipates to what the user wants or needs.

Context-aware: AmI systems can recognize specific users and its situational context and can adjust to the user and context. It may know that some users like classical music and others like jazz. Or that some users prefer it to be warm in the house and others like it cool. To give a contemporary example, car navigation systems can adjust the level of their lighting when it becomes dark, and the user benefits from more light on the screen. The system 'knows' that its user benefits from more or less light in different contexts.

Novel human-technology interaction paradigms: AmI systems will utilize new kind of interfaces which should support more seamless user experience with products and services. It is assumed that interaction paradigms like speech or haptics could lead to the more intuitive or natural interfaces.

Not all features are equally present in all AmI systems. For example, some AmI systems are very personalized, anticipatory and adaptive whereas others are not. There is also some overlap between adaptive, personalized, anticipatory and context-aware. The adaptive and anticipatory aspects, for example, make sure that the system can be personalized.

4.3.3 Neuroelectronics

There are roughly three branches in neuroelectronics. Each branch uses different devices to interface with the brain, and each of these devices has different features. The first branch, neuroimaging, uses different techniques to extract information from the brain to diagnose disorders or to study the brain. The second branch, Brain Computer Interfaces (BCIs), uses invasive or non-invasive electrodes to extract information from the brain, not for diagnostic or research purposes, but to control external devices such as wheelchairs, computers or airplanes. And the third branch, electrical neural stimulation, uses invasive electrodes to send electrical signals to specific parts of the brain. The only defining feature these three branches have in

Table 4.2 Defining features of emerging technologies

Emerging technologies	Defining features
Affective computing	Perceiving emotions/affects
	Expressive behaviour by computers/artificial agents
	Emotional cognition
Ambient intelligence	Embedded
	Interconnected
	Adaptive
	Personalized
	Anticipatory
	Context-aware
	Novel human-technology interaction paradigms
Neuroelectronics	Interface electrical devices with the brain
	Extract information from the brain
	Send electrical signals to the brain
Shared features of emerging technologies	National interaction
	Direct link between humans and technology
	Detailed understanding of the user
	Pervasive and invisible
	Autonomy of technology (potentially) power over the user

common is that they all interface electrical devices with the brain, either to extract information from the brain or to send electrical signals to the brain.

4.3.4 Shared Features of Emerging ICTs

Studying these examples of emerging ICTs reveal some shared features among them. A number of general trends concerning the relationship between humans and technology are summarized in Table 4.2. As a general rule, ICTs in the intermediate future are expected to become invisible, either by shrinking or becoming embedded in background technologies. They are expected to become pervasive and increasingly aware of and capable of reacting to their environment. As a consequence, these ICTs are expected to become more and more autonomous, which means they can react independently to their environment and in particular to user requirements.

This type of direct and embedded interaction allows for and requires a much more detailed model of the user. Significant amounts of data on the user are required for proactive and anticipatory services. The quantity of data will increase but also the quality, for example including emotional or location data.

This implies a fundamental change of the way in which humans use these ICTs or interact with them. One central expectation is that interaction will become more natural and seamless. This can be done in many different ways, but one central one is the direct link between the user, in particular the user's brain, and the technology. The exchange of information between user and machine can go both ways, so that it may become more difficult to distinguish between user and technology, thus moving closer to the vision of a cyborg.

The cyborg-like nature of the human-machine system means that it will be more difficult to distinguish between therapeutic and other users, thus opening the field widely to human augmentation and enhancement. This, in conjunction with the direct interaction means that the emerging ICTs have the potential to significantly change human views of themselves and also wield power over individuals.

4.4 Implications of Emerging ICTs from a CCAT Perspective

In this section we will explore how the Critical Capability Approach of Technology can be used to discuss the implications of the emerging ICTs identified earlier in this chapter.

First of all, there is a high level vision in the EU and indeed in many countries in the rest of the world that technological development as outlined in before mentioned examples of emerging ICTs is a positive and desirable trend (European Commission 2009, 2010). These technologies fulfill a certain vision of the future held by some politicians and citizens. It is generally perceived as progress to have an environment embedded with underlying technologies that "know" us to such an extent that they can make decisions on our behalf and interact with us. From a CCAT perspective, the principle of human centered development rejects the idea that technologies are intrinsically or universally valuable. They should be evaluated in terms of their contribution to emancipation, or individual freedom (both well-being and agency). If we apply this principle, there seems to be a need to scrutinize the motivations behind the development of these technologies. Do they provide more comfort, make us "happier", fulfill our desire, or do they build towards a society that we have rational reasons to value? Are these technologies means to ends (such as emancipation) or are they pursued as ends in themselves?

It follows from the above that the principle of human diversity is very important. A great variety of applications can be developed on the basis of these emerging technologies, targeting different user groups. Therefore instead of making sweeping judgment on whether these technologies should be developed and adopted, it is more important to examine what capabilities (in a capability approach sense) can they generate, for which users? For example, applications of ambient intelligence could be very useful for certain groups of disabled people in specific environments, which could enhance their autonomy in the sense that they can carry out their daily activities with less dependence on other people. This is very different from the scenario of an elitist ultramodern life style relying on an omnipresent technological environment. It is also important to examine what conversion factors need to be in place for the technologies to generate the types of capabilities that are considered valuable.

For instance mass diffusion of an AI embedded environment may raises challenges in energy consumption, environment, legal oversight and regulation, and so on.

The principle of protecting human agency prioritizes individual autonomy and agency, in other words self-determination and the ability to make changes to the world. From this perspective, there is serious need to reflect upon the trend of higher level of autonomy in technology as revealed by the shared features of emerging ICTs mentioned above. What impact does it have on individuals' freedom to control his or her own life, to make decisions on the basis of free will, and to retain sufficient options in building a world of difference? The consequences of accepting higher autonomy of technology at a societal level could be much more severe and difficult to amend, particularly when such technologies have become black-boxed, pervasive and invisible. In other words, when they become part of the furniture, taken for granted, and permeating our lives unquestioned, societies may be locked into the trajectory. Not that the technologies themselves automatically have deterministic power over society, but that social norms and practices developed in the structuration processes over time involving simple-minded adoption of these technologies are likely to produce structuring effects.

This is why the last principle is imperative to allow all member of society to participate in the discourse of technological development, before the black-box is closed, to explore different possibilities for the future. In history technological development has often been driven by market forces, authority or elite groups. Most people are just passive recipients of technologies. In this sense, the research presented here is of great significance, namely, the exploration of ethical implications of emerging technologies. By engaging with such a discourse, societies have the opportunities to reflect upon the choice of technologies for the future, and perhaps exert more agency over their "destiny".

These principles provide a grid of analysis that allows for a prospective evaluation of technologies. To return to the examples of emerging ICTs we have given in Sect. 4.2, one can apply such criteria to the defining features of technologies to explore their benefits and downsides. This would require an in-depth discussion and analysis of each of the technologies and their aspects, which goes beyond the confines of this chapter. The following table gives a brief overview of some of the more salient questions arising from the technologies (Table 4.3).

The table indicates that the criteria developed above are applicable to each of the sample technologies. A more detailed analysis would be required to gain an in-depth understanding and proceed to a full evaluation. As a proof of concept for the present chapter it suffices, however, by showing the relevance of the criteria.

4.5 Conclusion

The chapter set out to develop a theoretical framework that can be applied to evaluate emerging ICTs. It developed the framework, here called the "Critical Capability Approach of Technology" by drawing from Sen's Capability Approach and the

Table 4.3 Application of CCAT criteria to emerging ICTs

	Affective computing	Ambient intelligence	Neuroelectronics
Human-centred development	How do emotions affect our freedoms and capabilities? In what way can emotional computing further such emotion-linked capabilities?	Which lack of freedom is addressed by the technology? How are different deprivations accommodated?	What possible uses are envisaged beyond the high profile therapeutic ones? Which view of humans does neuroelectronics portray?
Human diversity	Diversity of emotions. How can a model of emotions be built that is appropriate to different cultures and individual variations of emotions?	Should ambient intelligence be implemented universally? Under what circumstances and for which users do they really enhance capabilities?	Do we have a sufficient understanding of the human brain to be able to specify individual capabilities aimed at by neuroelectronics?
Protecting agency	Emotions are directly linked to individual identity and agency. Does manipulation of human emotions erodes human agency?	Ambient intelligence is meant to be adaptive, context-aware and proactive. How can such features be used to strengthen human agency instead of weaken it?	By directly manipulating the brain, neuroelectronics can provide new means of expression, but it can also limit humans' ability to act. Can we foresee/address this?
Democratic discourse	How are relevant stakeholders identified? Which problems are subject to democratic discourse by which stakeholders? How can this be embedded in technology life cycles? What institutional support is required?		

Critical Theory of Technology. Using these underlying theories, it was argued that CCAT allows the determination of specific principles that can be used to evaluate technologies: these are the principles of human-centered technological development, of human diversity, of protecting human agency and of democratic discourse. These principles were then applied to three examples of emerging ICTs, namely affective computing, ambient intelligence and neuroelectronics.

The chapter should be seen as a proof of concept. It has shown that the principles of the CCAT are applicable to the three sample technologies. They provide additional insight into the nature and desirability of these technologies. This type of evaluation can be of interest to industry looking to develop technologies, as well as to policy makers who are trying to determine which technologies to promote or to regulate. It is clear, however, that this is only a first step.

The criteria developed here would suggest that ongoing discussions need to be held as technologies are developed. The evaluation of new technologies has to be an ongoing process if it is to remain sensitive to technical and social developments. This is of course not a completely new idea. The field of technology assessment, which has been established for decades, has explored numerous approaches and methodologies in this respect (Grunwald 2009; Decker and Ladikas 2004). There seems to be a new dynamic in this field, which is sometimes linked to the concept of "responsible innovation" (Owen and Goldberg 2010; Kjolberg and Strand 2011). The current paper can be seen as an attempt to contribute to this novel stream of debate and provide a set of accessible and yet theoretically sound principles for the evaluation process that needs to accompany any technology assessment activity.

A further issue worth considering is that CCAT provides a way of thinking not only about individual technologies but also about the constitution of society that is conducive to such reflection. It was already indicated that there is doubt whether leaving technology development solely to market forces is going to be conducive to the well-being, agency, freedom or emancipation of society. The criterion of democratic discourse in particular points in this direction. It raises the question who should have a voice in the decision which technologies are developed for what purposes. As indicated above, this is a fundamental question and we do not pretend to have a simple answer to it. To some degree it touches on some of the foundational assumptions of our societies. Despite this complexity one can argue that this is a core question that a modern and technology-dependent world will need to come to grips with.

Finally, there is an overarching question behind the chapter that pervades both the capability and the critical approach. This is the question in which world we want to live and how we decide upon this. The chapter has argued that technologies are means, not ends, which then calls into question the construction and development of ends. Modern and pluralistic societies do not have the option of setting generally accepted standards of belief and behavior that apply to everyone. In the same way it will be impossible to define technologies or their applications in unambiguous ways. Western democracies rely on principles (e.g. constitutions) and processes (e.g. public debate, elections) to come to an agreement about desirable futures. This chapter argues that these principles and processes should be extended to technology and it provides a set of ideas on how emerging technologies can be treated and evaluated so that the outcome of technical development work contributes to a world we want to live in.

References

Arnold, M. (2003). On the phenomenology of technology: The "Janus-faces" of mobile phones. *Information and Organization, 13*(4), 231–256.

Brey, P. (2000). *Disclosive computer ethics: Exposure and evaluation of embedded normativity in computer technology*. CEPE2000 Computer Ethics: Philosophical Enquiry. Presented at the CEPE2000 Computer Ethics: Philosophical Enquiry, Dartmouth College, Hanover.

Brey, P. (2008). The technological construction of social power. *Social Epistemology, 22*(1), 71–95. doi:10.1080/02691720701773551.

Buchanan, R. (2010). Human dignity and human rights: Thoughts on the principles of human-centered design. *Design Issues, 17*(3), 35–39. doi:10.1162/074793601750357178.

Cadili, S., & Whitley, E. A. (2005). On the interpretative flexibility of hosted ERP systems. *The Journal of Strategic Information Systems, 14*(2), 167–195.

Collingridge, D. (1981). *The social control of technology*. London: Palgrave Macmillan.

Cooke, B., & Kothari, U. (Eds.). (2001). *Participation: The new tyranny?* London: Zed Books.

Crocker, D. A. (2008). *Ethics of global development: Agency, capability, and deliberative democracy*. Cambridge: Cambridge University Press.

Decker, M., & Ladikas, M. (Eds.). (2004). *Bridges between science, society and policy: Technology assessment – methods and impacts*. Dordrecht: Springer.

Deneulin, S. (2006). "Necessary thickening": Ricoeur's ethic of justice as a complement to Sen's capability approach. In S. Deneulin, M. Nebel, & N. Sagovsky (Eds.), *Transforming unjust structures: The capability approach*. Dordrecht: Springer.

Doherty, N. F., Coombs, C. R., & Loan-Clarke, J. (2006). A re-conceptualization of the interpretive flexibility of information technologies: Redressing the balance between the social and the technical. *European Journal of Information Systems, 15*(6), 569–582.

European Commission. (2006, November 24–25). *From science and society to science in society: Towards a framework for 'co-operative research' – Report of a European Commission Workshop*. Governance and Scientific Advice Unit of DG RTD, Directorate C2. Brussels: European Commission. Retrieved from http://ec.europa.eu/research/science-society/pdf/goverscience_final_report_en.pdf

European Commission. (2009). *Preparing Europe for a new renaissance – A strategic view of the European Research Area – First Report of the European Research Area Board*. Brussels: European Commission. Retrieved from http://ec.europa.eu/research/erab/publications_en.html

European Commission. (2010). *COM(2010) 2020: Europe 2020 – A strategy for smart, sustainable and inclusive growth*. Brussels: European Commission. Retrieved from http://ec.europa.eu/eu2020/

Feenberg, A. (1999). *Questioning technology* (1st ed.). New York: Routledge.

Foucault, M. (1980). In C. Gordon (Ed.), *Power/knowledge: Selected interviews and other writings (1972–1977)*. London: Harvester.

Genus, A., & Coles, A. (2005). On constructive technology assessment and limitations on public participation in technology assessment. *Technology Analysis & Strategic Management, 17*(4), 433–443. doi:10.1080/09537320500357251.

Grunwald, A. (2009). Technology assessment: Concept and methods. In D. M. Gabbay, A. W. M. Meijers, J. Woods, & P. Thagard (Eds.), *Philosophy of technology and engineering sciences* (Vol. 9, pp. 1103–1146). Amsterdam: North Holland.

Hirschheim, R. A., Heinz, K. K., & Kalle L. (1995). *Information systems development and data modeling: Conceptual and philosophical foundations*. Cambridge: Cambridge University Press.

Introna, L. D. (2005). Disclosive ethics and information technology: Disclosing facial recognition systems. *Ethics and Information Technology, 7*(2), 75–86.

Kjolberg, K. L., & Strand, R. (2011). Conversations about responsible nanoresearch. *NanoEthics, 5*(1), 1–15.

Klein, H. K., & Myers, M. D. (1999). A set of principles for conducting and evaluating interpretive field studies in information systems. *MIS Quarterly, 23*(1), 67–93.

Kothari, U. (2001). Power, knowledge and social control in participatory development. In B. Cooke & U. Kothari (Eds.), *Participation: The new tyranny?* (pp. 139–152). London: Zed Books.

McAulay, L., Doherty, N., & Keval, N. (2002). The stakeholder dimension in information systems evaluation. *Journal of Information Technology, 17*(4), 241–255.

Ngwenyama, O. K., & Lee, A. S. (1997). Communication richness in electronic mail: Critical social theory and the contextuality of meaning. *MIS Quarterly, 21*(2), 145–167.

Oosterlaken, I. (2009). Design for development: A capability approach. *Design Issues, 25*(4), 91–102. doi:10.1162/desi.2009.25.4.91.

Orlikowski, W. J. (2007). Sociomaterial practices: Exploring technology at work. *Organization Studies, 28*(9), 1435–1448.

Owen, R., & Goldberg, N. (2010). Responsible innovation: A pilot study with the U.K. Engineering and Physical Sciences Research Council. *Risk Analysis: An International Journal, 30*(11), 1699–1707. doi:10.1111/j.1539-6924.2010.01517.x.

Peter, F. (2003). Gender and the foundations of social choice: The role of situated agency. *Feminist Economics, 9*(2–3), 13–32.

Robeyns, I. (2005). The capability approach: A theoretical survey. *Journal of Human Development, 6*(1), 93–114.
Robeyns, I. (2006). The capability approach in practice. *The Journal of Political Philosophy, 4*(3), 351–376.
Saravanamuthu, K. (2002). Information technology and ideology. *Journal of Information Technology, 17,* 79–87.
Sen, A. (1985). Well-being, agency and freedom: The Dewey Lectures 1984. *The Journal of Philosophy, LXXXII*(4), 169–221.
Sen, A. (1992). *Inequality reexamined.* Oxford: Oxford University Press.
Sen, A. (1993). Capability and well-being. In *The quality of life.* Oxford: Clarendon.
Sen, A. (1999). *Development as freedom.* New York: Knopf.
Stahl, B. (2008a). *Information systems: Critical perspectives* (Routledge Studies in Organisation and Systems). New York: Routledge.
Stahl, B. C. (2008b). Forensic computing in the workplace: Hegemony, ideology, and the perfect panopticon? *Journal of Workplace Rights, 13*(2), 167–183. doi:10.2190/WR.13.2.e.
Tao, J., & Tan, T. (2005). Affective computing: A review. *Lecture Notes in Computer Science, 3784,* 981–995.
van den Hoven, J. (2008). Moral methodology and information technology. In K. Himma & H. Tavani (Eds.), *The handbook of information and computer ethics* (pp. 49–68). Hoboken, N.J.: Wiley.
Zheng, Y. (2009). Different spaces for e-development: What can we learn from the capability approach. *Information Technology for Development, 15*(2), 66–82.
Zheng, Y., & Stahl, B. C. (2011). Technology, capabilities and critical perspectives: What can critical theory contribute to Sen's capability approach? *Ethics and Information Technology.* doi:10.1007/s10676-011-9264-8.

Chapter 5
"How I Learned to Love the Robot": Capabilities, Information Technologies, and Elderly Care

Mark Coeckelbergh

5.1 Introduction

Information technologies seem promising when it comes to improving care for elderly people. Intelligent systems could be used for the purpose of monitoring, care, and therapy. For example, robotic devices could assist elderly people to move around, autonomous pet robots could act as their artificial companions and allow therapeutic interaction, and telemonitoring systems (perhaps involving chip implants or intelligent nanobots) could allow medical professionals to keep track of people's health condition. In general, it seems that information technologies would allow people to live longer independently and better in their own homes.

In societies that face an ageing population, shortage of care workers, and political pressure to limit the allocation of public resources to health care, this may seem an attractive path. However, there are also potential ethical problems with this technological solution. For instance, some worry that privacy will be violated if data are sent to all kinds of other parties, that isolated living in a digitally enhanced environment would be deprive people of human contact and love, or that only the rich will enjoy the benefits of these technologies. For example, Sparrow and Sparrow have argued that the introduction of robots in the aged-care sector would be detrimental to elderly people's well-being since it would most likely result in a decrease in the amount of human contact (Sparrow and Sparrow 2006).

In this chapter I will not directly discuss these ethical issues, but sketch a general framework for the evaluation of information technologies in elderly care and illustrate its potential. I will first show that the capability approach is a useful and

M. Coeckelbergh (✉)
Department of Philosophy, University of Twente, P.O. Box 217, 7500 AE Enschede, The Netherlands
e-mail: m.coeckelbergh@utwente.nl

attractive way to make explicit what is at stake, ethically speaking, in elderly care. I will focus on Nussbaum's version of the capability approach and offer my interpretation of her list of capabilities as a list of criteria to evaluate the quality of lives – including the lives of the elderly. Then I will propose a modification to the capability approach that moves beyond a purely instrumentalist view of technology: drawing on earlier work I will argue that technology is not a mere means to realize capabilities, but that the criteria or goals themselves – the capabilities – change as a result of technology. Finally, I will further suggest methodological recommendations based on these insights and illustrate what this approach may mean in relation to elderly care by exploring a scenario of robotic elderly care in which information technologies transform people's capabilities for social affiliation and engagement in relations with humans and non-humans. In my conclusion I will indicate how this proposed interpretation of the capability approach is compatible with Nussbaum's ideas on the 'multiple realizability' of capabilities and the importance of moral imagination and moral sensitivity.

5.2 A Capability Approach to Ethics of Care Technology

One way to articulate both the promises and the ethical worries concerning the use of information technology in elderly care is to use the capabilities approach as a descriptive and evaluative framework. The capabilities approach was initially developed by Sen and Nussbaum in response to standard approaches to human development in development economics. They argued that well-being should not be evaluated by looking at the Gross National Product (GNP) of a nation, or, more generally, at the material resources people have; instead, they argue, we should focus on what people are actually able to do (Nussbaum and Sen 1993; Nussbaum 2000), on expanding 'the real freedoms that people enjoy' (Sen 1999, p. 3), and, more fundamentally, on their human dignity (Nussbaum 2006). Human development – in 'development' countries and elsewhere – is not (only) about giving people formal freedoms, material goods, and technology; it is about empowering people to live better lives.

I propose that we apply this approach to ethics of elderly care.[1] It is not enough to give elderly people care technology; what matters is what these people can do with it in relation to their well-being, that is, what matters is that it *enhances* their well-being and agency. From this perspective, the promise and goal of using information technology in elderly care can be framed as empowering people to live independently, to enjoy a higher quality of life, and to live their lives in dignity. This puts the emphasis on what people can do with the technology (the goal) rather

---

[1] I limit the topic of this chapter to ethics of elderly care but the capability approach can be more widely applied across various domains of ethics and ethics of technology.

than on the technology itself and its particular technical details. There is no point in giving people the most advanced technological equipment if they cannot use it to improve the quality of their lives. More, it is ethically unacceptable if it diminishes their human dignity.

In order to evaluate whether or not a particular technology actually contributes to that quality of life and indeed to elderly people's human dignity (the ultimate end Nussbaum puts forward), we may use Nussbaum's list of capabilities as criteria to evaluate the quality and dignity in elderly care.[2] The list includes the following 'central capabilities':

1. Life: 'Being able to live to the end of a human life of normal length; not dying prematurely, or before one's life is so reduced as to be not worth living.'
2. Bodily health, including nourishment and shelter
3. Bodily integrity: free movement, freedom from sexual assault and violence, having opportunities for sexual satisfaction
4. Being able to use your senses, imagination, and thought; experiencing and producing culture, freedom of expression and freedom of religion
5. Emotions: being able to have attachments to things and people
6. Practical reason: being able to form a conception of the good and engage in critical reflection about the planning of one's life
7. Affiliation: being able to live with and toward others, imagine the other, and respect the other
8. Other species: being able to live with concern to animals, plants and nature
9. Play: being able to laugh, to play, to enjoy recreational activities
10. Control over one's environment: political choice and participation, being able to hold property, being able to work as a human being in mutual recognition

(Nussbaum 2006, pp. 76–78; my summary)

This list of capabilities can be used as a conceptual tool to make explicit what is at stake, ethically speaking, in elderly care and to develop criteria to evaluate the quality of that care.

First, while such general criteria may not allow us to solve particular difficult cases or to determine the outcome of practical ethical deliberations, they can guide our reflection as ethical signposts in the following way. If what matters is that elderly people can live the last years of their lives in dignity, then the list makes explicit that such dignity means and what good elderly care should aim for. This helps us to speculate about what technology might do to people and to ask the right questions. For instance, using the capabilities as criteria for evaluation we may ask: Does the technology really enhance the capability of affiliation with others or does it only allow us to 'stay connected' while diminishing real human contact? And if intelligent systems were to take over some decisions, would they sufficiently respect people's own capability of practical reason? Would bodily integrity be respected if intelligent nanobots were to 'live' in the body? Moreover, since the capabilities approach has

---

[2] This argument can also be applied to health care in general (Coeckelbergh 2010a, b).

always been concerned with issues of justice, one could ask if these technologies will only benefit elderly people in technologically advanced countries and if that is problematic from a social justice perspective. There may be gaps within one country – between people who can afford the high-tech care and people who can't – and between 'advanced' and 'developing' nations.

Second, this focus on capabilities does not only allow us to specify how the technology might change people's lives and to ask ethical questions; if further developed it also enables us to *evaluate* these changes, to answer the questions. Starting from Nussbaum's list, one can set more concrete thresholds that specify which level of which capability must be reached. For instance, one could specify a minimum level of physical health, below of which the life of the elderly person is judged to be 'not worth living' (the first capability of Nussbaum's list), and aim for bringing every elderly person over the threshold.[3] The list leaves open what this threshold should be. This is to be regarded as an advantage. The capability approach, with its stress on dignity and its list-approach, is 'universalist' enough, yet at the same time it also leaves plenty of room for interpretation in particular contexts. According to Nussbaum, the list can be 'specified in accordance with local beliefs and circumstances' and the threshold level needs to be determined by political consensus (Nussbaum 2000, p. 77).

Moreover, as a capability approach, this approach to ethics of elderly care has the further advantage that it moves beyond the abstract discourse of liberal rights in so far as these rights give people only formal freedoms and protections. For example, we agree that all elderly people should have the right to life and liberty, as the Universal Declaration of Human Rights prescribes. But this says little about what elderly people can actually do given their specific condition (physical condition, dependence on others, etc.). Using the capability approach, we can analyze and evaluate elderly people's capabilities given their specific conditions and in particular contexts and circumstances. With a view of evaluating information technology, then, we can analyze how the specific technology changes or might change elderly people's capabilities and using thresholds (agreed upon) we can evaluate these changes. For example, *if* in a particular home care context artificially intelligent information technology were to diminish elderly people's capability for social affiliation to a degree that is judged to be ethically unacceptable, then the use of this technology in elderly care would be unacceptable.

But what is social affiliation in this context, and can it be defined independently from the technology? Can *any* capability in elderly care be defined independently from technology? What do we mean by 'technology'?

---

[3] Note that judging a particular life to be 'not worth living' does not necessarily imply that it is therefore justified to end such a life (e.g. suicide or euthanasia). On the contrary, it seems that a capability approach would rather require us to help the person to reach the minimum level. And if this were impossible, then there may be other ethical considerations for blocking the suicide or euthanasia option.

5.3 Towards a Non-instrumentalist View of Care Technology

Proponents of the capability approach, if concerned with technology at all, tend to view technology as a mere means to its ends – ends formulated in terms of capabilities and their ultimate aim (e.g. human dignity or freedom). Applied to elderly care as proposed in the previous pages, a capability approach that were to keep this assumption would see technology as an instrument to realize the aims of elderly care, here expressed in terms of capabilities. Hence little attention is paid to the technologies themselves and what they do to the very meaning of the capabilities we wish to use as evaluative criteria.

One specific reason why capability theorists inexplicitly assume this instrumental role of technology (the more general reason is too little attention to technologies and their unintended effects, a lacuna shared with most other theories in practical philosophy), lies at the heart of the 'classic' version of the capability approach as explained above, which says that it is not enough to give people material *means* if with this means the goals of development – e.g. human dignity (the *end*) – are not reached. Thus, the very motivation for people in development studies to adopt a capability approach can be framed as a move from means to ends: focus on which end people can reach rather than on the means (money, material goods, technology).

However useful this shift has been for moving forward development studies, it is worth considering the limitations of thinking along the lines of this means-ends scheme. These limitations are at least partly due to a misguided understanding of the relation between technology and human ends (e.g. ethical values). Contemporary philosophy of technology teaches that technologies are *not* mere means but 'do' more than we intend to do with them. They are not mere instruments or tools, but change our goals and what we consider to be important. Consider for example how modern means of transportation have changed the way we organize work and leisure time, how the anti-conception pill has changed how we shape and evaluate personal relations, and how mass communication media influence our life projects. The goals of elderly care – whether or not we articulate them by using the capability approach – are not immune for such techno-moral change. For example, today people around 65 have different lives, but also different expectations and aspirations due to the exponential development of medical technology during the past century.

If this is true, we can no longer take for granted that what we now or in the past meant by health, affiliation, etc. will remain the same in the future. This renders it difficult to evaluate the promises of information technology for elderly care in the future. One answer to this difficulty is to recommend using our moral imagination. I will say more about this below. However, based on current transformations, we can already discuss – and indeed *question* – some (current) ethical worries with regard to information technology in elderly care. For instance, if we are concerned about privacy, we need to take into account that information-technological practices today do not promote privacy and that many young people care less about privacy than people did in the past. Bodily integrity has already been 'violated' by many pharmaceutical products. And for many of us the capability of affiliation has been transformed by information technology – in particular mobile technologies.

Indeed, the elderly we are concerned about in the scenario above are not grandma or grandpa, but you and I. Our capabilities have already been transformed by information technology and will continue to be transformed until we reach the point that we need (more) elderly care. The meanings of health, affiliation, play, etc., therefore, are not independent from information technologies but are partly constituted by these technologies. They change as technology changes.

Acknowledging this weakening of the means-ends dichotomy does not imply that all our ethical worries are necessarily misguided, but it means that we have to be far more precise when we voice them. For instance, if we care about friendship and love and claim that information technology threatens the possibility for realizing and enjoying friendship and love, then we have to show why the use of particular information technologies like software for social networking and on-line communication, which are likely to be used by the elderly of the future (and are used already today), would be harmful for the kind of relations we value most. But in addition we also have to consider that what we mean by friendship and love might be and become influenced by the very technologies under discussion.

5.4 Imagining and Interpreting Capabilities: A Scenario of Robotic Elderly Care

Such an exercise cannot be done without engaging with current and emerging technologies, and with users and designers of the technologies. 'Emerging' is important here: we do not only want to know what current technologies do to our capabilities: we also want to know what *future* technologies could do to them.

Whether or not we can obtain knowledge about the future is a perennial philosophical issue and I shall not provide a comprehensive discussion of it here. For the purpose of this brief inquiry, let me make explicit my own preferred approach, which can be summarized as the use of 'moral imagination', or, more precisely 'techno-moral imagination'. Although we can never obtain certainty about the future, we can explore future possibilities by using our imagination: by imagining technological change and its potential consequences for the moral life. This use of the imagination is not arbitrary, but is based on available information and is preferably active and creative. One way to proceed in this case is to study technological promises (research proposals, interviews in the media, and so on) and to write – not just read and discuss – fictional scenarios in order to imaginatively explore how future technologies could re-shape not only elderly care, but also our capabilities and their meaning.

Although fictional scenarios have been used before in exploring how emerging technologies bring about moral change (Swierstra et al. 2009),[4] the capability

---

[4] My use of a fictional scenario is inspired by my research on moral imagination (Coeckelbergh 2007) and by collaboration with Tsjalling Swierstra on moral change and techno-moral scenarios, in particular my experience of writing scenarios about nanotechnology for his 'Vignetten en scenario's' Nanopodium project (2009–2010).

approach offers a precise and helpful way to frame the ethical, normative issues at stake. Consider the following scenario, which explores how information technologies might re-shape elderly people's capabilities for social affiliation and engagement in relations with humans and non-humans.

> January 26, 2060. Grandpa wakes up in his CareCap.[5] Robodog Simply jumps on his bed and greets him – he's clearly happy to see that his companion is awake and appears to feel better. The friendly robot helps with washing, plays a little bit with him, and then fetches breakfast for him. Yesterday was a bad day: Simply had to do a medical intervention when the NanoCare system detected problems in the belly. But now grandpa is ready again for a chat and a game. This will not only keep his brain functions going; he will also have fun and improve his skills. Via the net he talks with his friends and plays a game with his grandchildren.
>
> Grandpa remembers that in the old days, when he was still working (there used to be a sharp division between working life and pension), he regarded the idea of 'isolated' e-care as a nightmare idea: he thought it would be utterly inhumane to lock up people in a high-tech environment with robots and other electronic stuff that provides 'care'. But when he considers his condition now, he has little to worry about. It turned out that his elderly life in CareCap was not so different from his working life towards the end of his career: at the time he was also always connected to the net, his health was monitored by RFIDs, and he spent more time talking to his children on the mobile phone and via FaceNet than seeing them in 'real' life. What was real? It was true that he had now less physical contact with other people, and was more 'isolated' in this sense, but it was also 'real' that via the net he was connected to the whole world, including people in the same situation and condition. And a robotic dog was much more intelligent and pleasant than a real one (in addition to the fact that it was much easier to maintain). What was freedom? With his bodily condition, he could not live long outside his capsule, true. He needed the electronic support systems and could only make short robot-assisted excursions into the green area (not the red!). It is true that he had not the freedom to travel with his body. But his mind was free to dwell in all 'real' and fictional places via the net. Via FaceNet and TwitImage many people knew about his activities. He had plenty of followers. In a sense he felt even more active than when he was a child. His mental life was surely more interesting than that of his parents when they were at the height of their capacities. His body too was much healthier now, due to the different care systems. And he had great fun with the robot dog. For sure, he thought, I'm more happy than they were in their old age, in their homes for elderly care, where they had 'human contact' but where their minds and bodies were slowly withering away.

With this fictional scenario, I do not wish to promote this particular kind of technologies, elderly care and indeed this conception of good elderly care and the good life. But it serves as an illustration of what I have in mind when I claim that the relation between technology and capabilities is not merely instrumental. The scenario suggests that in the future we might interpret capabilities differently, due to changes to technologies and to our lives. Of course we may disagree with this particular interpretation. We might want to further discuss it. But in any case the meaning of capabilities is not fixed but requires a hermeneutic process, a work of interpretation, which can leave out technology only at the expense of distorting our understanding of the practice of care – and indeed any practice. When we think of social affiliation,

---

[5] Care Capsule: semi-closed system for habitation, monitoring, and care of elderly people over 100.

for instance, we usually construct the meaning of this capability without referring to technology, for example by referring to experiences of two physical people – humans – interacting in a physical space. We might think of companionship, friendship, or love between two humans in a natural environment. But today as in the past, such imagery of social affiliation does no justice to the nature and plurality of human experience. Social affiliation takes and has always taken different forms and has often involved technology in various ways. Consider architecture, music, letter writing, and other 'humanistic' and 'cultural' yet highly technologically mediated faces of social affiliation. Affiliation between humans is often directly and otherwise indirectly co-shaped by technologies. Moreover, to restrict social affiliation to affiliation between humans does no justice to the extent, significance, and variety of relations between humans and non-humans, such as pets or – perhaps for some of us – robots. It may be that these non-humans are 'mere animals' or 'mere machines', but as a matter of fact in many contexts we do not treat them as such and act as if they belong to our social world. Hence, the capability of social affiliation should not be interpreted as excluding such relations a priori, but should be related to other capabilities Nussbaum mentions: relations to other species, play, etc. In this sense, there is not so much a 'list' of capabilities but a normative-hermeneutic 'web' of capabilities.

Applied to the question regarding elderly care, capabilities, and information technology, this approach suggests that an evaluation of information technology in this context should not be restricted to an evaluation *of* the technology *in the light of* capabilities, since that formula assumes a purely instrumental relation between the 'object' of evaluation and the evaluative 'criteria'. Instead, the evaluation should involve a discussion of the capability and its relation to technology. The fictional scenario explores just one way in which not only the practice of elderly care but also the very meaning of the capability of affiliation might change due to information technology. It suggests that future technology – but also the way technology shapes our lives today – is not only likely to change our lives, but also how we evaluate those lives: it is likely to alter what we consider a good life, dignity, freedom, and good social affiliation. It transforms the criteria of evaluation as much as it transforms us.

Thus, whether or not we wish to endorse the particular vision of technology and of the techno-care future suggested in this fictional scenario, the exercise reminds us that we should remain critical of the instruments we use to judge whether or not this vision counts as a beautiful dream or as a nightmare. Here it means that if and when we use the capability approach to evaluate the use of information technology in health care, we should carefully work out not only what we mean by 'social affiliation', 'bodily health', 'control over one's environment' etc., but also how the meaning of these terms might change in the future as a result of the very technologies we are trying to evaluate.

Note that this does not 'weaken' the capability approach as a critical tool. I concede that by connecting the concepts 'capability' and 'technology' in this way, this interpretation of the capability approach reduces the distance between principle and practice, between evaluative criteria and objection of evaluation. It seems to reduce the normative power and authority of the criteria themselves. In this sense the approach may become less 'solid' as a normative framework: it seems to be built on criteria that are unstable (capabilities). However, at the same time the ethical-hermeneutical interpretation of the capability approach proposed here opens up a second space from which normative power may be derived: by using our moral imagination, we can create a space between the present and the future (or the past), which allows us to achieve critical distance from our current ethical points of view. In other words, it adds another, less principled and more historical-imaginative pathway that may help us to achieve what Socrates considered to be the aim of ethics.

5.5 Conclusion

The capabilities approach offers a powerful conceptual tool for gaining more insight in the ethical issues concerning the use of information technology in elderly care, in particular if it is interpreted and modified in the way proposed above and if it is combined with well-informed moral-imaginative work. Using the capability of 'social affiliation' as an example, I have explored a fictional scenario in order to illustrate how this application and modification of the capability approach requires us to reflect on what information technology might do to elderly care *and* to the criteria we employ to evaluate those possible changes. This ethical-hermeneutical interpretation and use of the capability approach may make it a less 'solid' (if that should be a goal in normative ethics at all), but more adequate conceptual instrument to capture and explore what information technologies might do to our lives – now and in the future.

For those of us who feel attracted to Nussbaum's version of the capability approach, this interpretation is good news for at least two reasons. First, the proposed hermeneutical approach is compatible with what Nussbaum (perhaps somewhat misleadingly[6]) calls the 'multiple realizability' of capabilities: every society needs to discuss what a particular capability means for a particular society. Items on the list 'can be more concretely specified in accordance with local beliefs and circumstances', which leaves room for political deliberation (Nussbaum 2000, p. 77). What my proposal implies and emphasizes, is that this dialogical and interpretative process has to be re-done in the light of new technologies. In other words, interpretations of capabilities differ not only in place (different cultures), as Nussbaum has argued, but also in time, for example when new media and technologies change our ideas and our practices. Second, Nussbaum has always recommended the exercise of moral imagination and fiction as a means to explore different lives and possibilities

---

[6] The term 'realizability' suggests that the capabilities and their meaning are fixed, whereas there might be different interpretations about how to realize them in practice. This may be right, but in the light of the ideas presented in this chapter, I prefer to interpret Nussbaum on this point as meaning that the capabilities themselves are also up to interpretation (and indeed negotiation).

in order to become more morally sensitive and practically wise. Indeed, the exercises of techno-moral imagination recommended here are not 'mere science-fiction' if that means they are both unreflective and irrelevant to important and pressing ethical concerns. If there is anything that can save elderly care from a dark and cold future, it is critical reflection that is not only motivated *but also deeply shaped* by our empathy and sympathy with future generations, with less well-off people, with our children and our parents, and indeed with our near-future elderly selves. After all, the question concerning technology and elderly care is *our* question.

References

Coeckelbergh, M. (2007). *Imagination and principles*. Basingstoke/New York: Palgrave Macmillan.
Coeckelbergh, M. (2010a). Human development or human enhancement? A methodological reflection on capabilities and the evaluation of information technologies. *Ethics and Information Technology, 13*(2), 81–92.
Coeckelbergh, M. (2010b). Health care, capabilities, and AI assistive technologies. *Ethical Theory and Moral Practice, 13*(2), 181–190.
Nussbaum, M. C. (2000). *Women and human development: The capabilities approach*. Cambridge: Cambridge University Press.
Nussbaum, M. C. (2006). *Frontiers of justice: Disability, nationality, species membership*. Cambridge/London: The Belknap Press of Harvard University Press.
Nussbaum, M. C., & Sen, A. (1993). *The quality of life*. Oxford: Clarendon.
Sen, A. (1999). *Development as freedom*. Oxford/New York: Oxford University Press.
Sparrow, R., & Sparrow, L. (2006). In the hands of machines? The future of aged care. *Minds and Machines, 16*(2), 141–161.
Swierstra, T., Stemerding, D., & Boenink, M. (2009). Exploring techno-moral change: The case of the obesity pill. In P. Sollie & M. Düwell (Eds.), *Evaluating new technologies* (pp. 119–138). Dordrecht: Springer.

Chapter 6
Towards a Sustainable Synergy: End-Use Energy Planning, *Development as Freedom*, Inclusive Institutions and Democratic Technics

Manu V. Mathai

> *Some of our mentality about what it means to have a good life is, I think, not going to help us in the next 50 years. We have to think through how to choose a meaningful life where we're helping one another in ways that really help the Earth.*—Elinor Ostrom (2010)
>
> *Focusing on human freedoms contrasts with narrower views of development, such as identifying development with the growth of gross national product, or with the rise in personal incomes, or with industrialization, or with technological advance, or with social modernization.*—Amartya Sen (1999: 3)
>
> *It is by institutional extension that subjective impulses cease to be private, willful, contradictory, and ineffectual, and so become capable of bringing about large social change.*—Lewis Mumford (1970: 424)

6.1 Introduction

Today's environmental crisis is characterized by scales of energy and material throughput that are not ecologically viable; a socially unequal distribution of the impacts of that throughput; and by a persistence of this throughput and inequality despite innovation, effort and "success" in addressing them over nearly four decades. Nearly 60% of our planet's ecosystem services, which were subjected to scientific scrutiny, demonstrates declines in viability. At the local level, clean air, fresh water and nurturing land remain a privilege for an affluent minority despite the arrival of the "gilded age" of modernity (IPCC 2007; Millennium Ecosystem Assessment 2005).

M.V. Mathai (✉)
Institute of Advanced Studies, 6F, International Organizations Center,
United Nations University, Pacifico-Yokohama, 1-1-1 Minato Mirai,
Nishi-ku, Yokohama 220-8502, Japan
e-mail: manu@udel.edu

By and large institutional responses to this crisis assert faster economic growth and better management of the environment through "improved" technologies and economic and regulatory instruments (Munasinghe and Swart 2005; WCED 1987). The recent revival of civilian nuclear power in India (and elsewhere) is an illustration of this. It is seen as crucial for sustained 8–10% GDP growth rates while limiting GHG emissions. Crucially, this approach proceeds without questioning modern society's easy faith in its scientific and technological means, while reflection on and engagement with the question of ends of growth and development remain sidelined.

Yet, such complacency is no longer viable. Over the decades, despite strong economic growth, efficiency gains through scientific and technological advances and better management, the persistence of the environmental crisis and its further intensification is acknowledged. The solution to pursue faster economic growth to realize higher per capita income, so that affluence purportedly will hedge against and/or reverse the crisis is challenged at two fundamental levels. The promise of the environmental Kuznets curve[1] that undergirds this thinking remains limited and inconclusive (Dinda 2004) and concerns raised about the finitude of the earth's ability to furnish useful matter and energy and absorb waste remains logically and empirically sound (Daly 1990, 1991, 1996). Although innovation and advances in efficiency and productivity can be helpful, they are insufficient to realize sustainable scales of material and energy throughput (Herring and Roy 2007; Byrne and Toly 2006; Haberl et al. 2006; Wilhite and Norgard 2004; Sachs 2002; Bunker 1996). Finally, given the bureaucratic tendency to exclude people, caution is urged regarding the ability for better management to respond to the crisis (Norgaard 1994; Sachs 1992; Sen 1999, 1995a, b; Habermas 1973).

Despite such limitations, mainstream public policy negotiating the environment-development issues remains based on conventional strategies. Inquiring into why this is the case and using civilian nuclear power in India as an illustration, we have argued elsewhere (Mathai 2010) that this status quo derives from a dominant economic development discourse of "Cornucopianism" complemented structurally by an approximation of a "megamachine[2] organization of society." Cornucopianism is understood as combining modernity's uncritical proclivity for advancing scientific and technological means and its liberal democratic ideal that social justice is predicated on economic growth and expansion (Byrne and Yun 1999). The megamachine organization of society is a metaphor for the technological-institutional constellation that embodies this discourse. It privileges concentration of executive power, accords primacy to big science and technology, is invested in commodification and an ever expanding capitalistic economic order, and overall a situation of "quantification without qualification" (Mumford 1970; Mathai 2010). This dominant overall view

---

[1] The environmental Kuznets curve, widely known as the "inverted U-curve," prognosticates that environmental impact, plotted on the y-axis, grows as per capita income grows, then plateaus and decreases as per-capita income, plotted on the x-axis, continues to grow. The upshot of this purported relationship is that growing per-capita income will save the day, or as John Tierney (2009) puts it, "Use Energy, Get Rich and Save the Planet"!

[2] The "megamachine" is a metaphor honed by Lewis Mumford through his studies of the history of technology and inquiry into the Manhattan Project. See Mumford (1970).

on science, technology and society has a number of implications in the context of economic development and energy policy. Firstly, it tends to presume that overabundance is an imperative for viable social and economic development. This course is futile given the pressing need to find sufficiency. Secondly, it is predisposed to centralization and as it skews deliberation on energy-society relations to technocratic terms it restrains deliberation to venues limited to the ruling, technocratic elite (Mathai 2010).

What appears to be needed to foster sustainable and equitable energy and material throughput are reductions, immediately in the case of industrialized countries, and a period of *equitably distributed* growth followed by stabilization, in less industrialized countries. This is a call for efficiency as well as sufficiency where the latter is a distinctly non-technological undertaking. If these arguments are valid, then what are the alternative responses to a crisis driven by unsustainable scale and inequity? As an exercise in exploring alternatives, this chapter enjoys the benefit of drawing on distinguished and creative lines of thought. In place of Cornucopianism this chapter inserts the discourse of "ends" and "human freedoms" and in place of the megamachine organization it offers "sustainable structures of living together" and "democratic technics." It proceeds by introducing and discussing strategies for these goals before considering the synergy between them. The alternatives discussed include the "Development Focused End Use Oriented and Service Directed" (DEFENDUS) approach for energy planning; the Human Development and Capability Approach and the Sustainable Energy Utility as an institutional template. It concludes by brining attention to the technological infrastructure – democratic technics – considered essential for the synergy being proposed here.

6.2 Toward a Human-Centered Development, Energy and Environment Discourse

This proposal to revive a human-centered orientation in public affairs calls for strengthening society's ability to take stock of its present, situate itself in the currents of its history and resolve toward a particular future. In this effort, it is imperative that the complete human personality,[3] which is the only locus of a full social consciousness, be given far greater space than the routine reproduction of economic development, technology innovation and energy policy experience is capable of.

Removed from the fold of mainstream environmentalism, considerable work exists that moves in such a direction. Typically, this work accords heightened attention to the normative question of "ends." With reference to energy policy, the work

---

[3] This image as conjured by Lewis Mumford (see Mumford 2000, 1963 and others), captures the confluence and complimentarity between the objective and subjective sides of the human personality. It attends to material necessities of life as well as, with equal dexterity, to the emotional and normative necessities. However, in light of the modern, one-sided valorization of instrumental values in the development discourse, the complete human personality is often backstaged and the narrow personality type, represented by categories such as the technocrat or the economic man dominate.

carried out by the self-described "group of four" is of much importance. Introducing their book Energy for a Sustainable World (Goldemberg et al. 1988) these authors note they are:

> especially interested in understanding how patterns of energy use might be shaped so as to promote the achievement of certain basic societal goals, equity, economic efficiency, environmental soundness, long-term viability, self-reliance and peace.

Further, these authors propose that the basis for formulating energy strategies compatible with the above goals comes from "shifting the focus of energy analysis from the traditional preoccupation with energy supplies to the end-uses of energy" (Goldemberg et al. 1988: 3). One member of this group, Prof. Amulya Reddy, was instrumental in extending this line of thinking to the energy-development context of developing countries by proposing and applying an energy planning framework referred to as DEFENDUS (Development Focused, END-Use oriented and Service directed), which is discussed in the next section.

In the field of economic development, insightful theorization and research has proceeded on the question of "means" and "ends" in the context of human well-being. Widely considered as pioneers in this area, Amartya Sen and Martha Naussbaum's work is being developed and extended by a community of researchers and activists working with their "Human Development and Capability Approach" (hereafter "capability approach"; Sen 1999; Nussbaum 2000; Alkire 2002; Stewart and Deneulin 2002; Deneulin 2006; Comim et al. 2008; Deneulin and Shahani 2009, among others).

While both these efforts seek to radically reconceive energy and economic development policy, they tend to overlook extant political economy. As Stewart and Deneulin (2002: 64) note with regard to the capability approach, the problem is that Amartya Sen's "concept of democracy seems an idealistic one where political power, political economy, and struggle are absent." Similarly, we offer that the DEFENDUS approach might benefit from a clearer appreciation of the political economy of energy policy. In its present form, its end-use orientation is overwhelmed by existing institutional tendencies. Evidently then, applying the capability approach and DEFENDUS needs to face-off the predispositions of existing political economy. In this context we explore as a template an arrangement of energy-society relations developed by the University of Delaware Center for Energy and Environmental Policy – known as the *Sustainable Energy Utility* (SEU, SEU 2007; Byrne et al. 2009) – that seeks to circumscribe the hegemony of the extant energy arrangement.

6.3 End-Use Energy Planning: The "DEFENDUS" Framework

The DEFENDUS methodology evolved during the 1970s and 1980s, with its roots in the legendary collaboration between Jose Goldemberg, Thomas Johansson, Amulya Reddy and Robert Williams, which resulted in the publication of *Energy for a Sustainable World* (Goldemberg et al. 1988). Highlighting the basic idea guiding their collaboration, they note:

> We do not treat energy supply or consumption as an *end* in itself. Rather, we focus on the end-uses of energy and the services that energy performs. Energy use is, after all, *only a means* of providing illumination, heat, mechanical power, and the other energy services associated with satisfying human needs. Here we are especially interested in understanding how patterns of energy use might be shaped so as to promote the achievement of certain basic societal goals—equity, economic efficiency, environmental soundness, long-term viability, self-reliance, and peace. *Ours is, in short, a normative analysis.*

Urging such a distinction between "ends" and "means" is a significant change in priorities and values of the modern mind, generally, and specifically in its relationship with energy.

The more distinctive articulation of DEFENDUS, with a development focus and India orientation, followed in Reddy (1990) and Reddy et al. (1995a, b). It placed the energy system as a sub-system of the larger economic system and argued that the energy sub-system be directed by the considered and desired ends of the larger economic system:

> Systems involving human beings are goal-oriented, and the purpose of energy planning is to make the energy sub-system drive the goal-oriented system towards its goal(s). Every goal implies choices, values and preferences, and therefore *a goal-oriented approach is a normative approach that defines what is desirable.* (Reddy et al. 1995a: 17, italics added)

The framework offers two critical "controls" in the energy planning exercise. First, its "goal orientation" allows the planner to express values of the community for whom energy planning is being undertaken. It is said that this step involves the "heart." Second, the "end-use focused control" requires the planner to identify energy services aligned with the various end-uses that result from the chosen development focus and apply technical knowledge to minimize inefficiencies and harness renewable energy alternatives. It is said that this step involves the planners' "head" (Reddy 1990: 21).

In its application the DEFENDUS framework has three basic components: (a) a methodology for building "demand scenarios" for particular energy carriers or fuels; (b) "end-use orientation and direction towards energy services (*rather than consumption*)" (italics added); and (c) a methodology for determination of "least-cost supply mix" to meet the energy requirements of the selected demand scenario (Reddy et al. 1995a: 19). The "demand scenario", as the name suggests, is derived from scenarios of the future. It is different from projections, which relate the future to the present through (often arbitrarily selected) mathematical relationships. Thus the key question posed is: "if measures M1, M2, M3… are implemented, what will the result be?" These "measures" are derived from the "strategies" for the energy sub-system, which in turn are derived from the "goals" identified for the energy sub-system. In turn, the "goals" of the energy sub-system are guided by the task of accomplishing the ends valorized by the community in their scenario for the future (Reddy et al. 1995a: 19).

The second step is to undertake the "end-use" analysis. It brings "end-uses of energy and the services to be derived from energy" to center-stage (Reddy et al. 1995a: 19). This step is based on the understanding that "technological improvements can lower the need for energy while retaining the same level of energy derived services" (Reddy et al. 1995a: 19). As such all end-uses are analyzed with two specific goals in mind. First, can the same level of service be delivered by a more efficient technology? An example is that of CFL bulbs and their ability to deliver the same illumination, for roughly one-fourth the electricity consumed. The second goal in end-use analysis is to explore possibilities of substituting conventional energy sources or carriers and architectures with renewable sources or more localized architectures, at the same or lower cost. An example in this regard is the substitution of electric water heaters by solar water heaters.

The third step is to make choices among different options. All available energy options, including generation, efficiency and substitution, are assembled on the basis of their levelized cost. Selection of technologies starts with the lowest cost energy option, and up the ladder until the cumulative energy savings, substitutions or generation from the selected options can meet the energy demand of the selected future scenario.

6.3.1 An Example of the Application of the DEFENDUS Approach

A famous application of the DEFENDUS approach was its use in 1987 to analyze electricity planning in the state of Karnataka in South India. Reddy (1990) notes that the conventional projections of the local Committee for the Long Range Planning of Power Projects (LRPPP) are based on a simple application of a compound rate of growth from the immediate past to base year electricity demand. Such "*business as usual* projections generally exclude the possibilities of improvements of energy efficiencies and alterations of growth rates, so that the future is viewed as an amplified version of the recent past" (Reddy et al. 1995a: 16, italics in original). To challenge the LRPPP, Reddy (1990) and Reddy et al. (1995b) applied the DEFENDUS approach to electricity planning for Karnataka for the same period (1986–1987 as the base year and 2000 as the "planning horizon").[4]

The "development focus" used to inform demand scenarios had three normatively justified ends. First, that "every single home will have an electric connection and electric lights instead of kerosene lamps;" second, prioritization of employment generation through the insight that non-power intensive industries create more jobs. This end entailed a faster growth rate for low-tension electric connections; third, the electrification of all agricultural groundwater pumps within the limits of the state's groundwater potential. Being a primarily agrarian economy (in terms of population employed) this was considered an important development focus.

---

[4] Also see Sant and Dixit (2000) for a DEFENDUS inspired analysis undertaken for the electricity sector of Maharashtra, which realizes significant financial and environmental benefits compared to the conventional electricity plan.

The electricity demand of this development focus was treated with the second step: end-use analysis. This process identified a number of possible efficiency improvements and substitutions. The measures included: replacement of incandescent bulbs with compact fluorescent bulbs; retrofitting irrigation pumps with frictionless foot-valves and use of low friction pipes to transport the water; industrial modernization with more efficient motors, furnaces and boilers and newer processes. Also identified were fuel and carrier substitution; solar-water heaters to replace electric water heating in middle-class urban homes and use of liquefied petroleum gas as a substitute for electric cooking in urban homes.

The results of this exercise undertaken in the late 1980s, with the year 2000 as the planning horizon, are spectacular. When compared to a projection of 9% growth (LRPPP) until the year 2000, the demand scenario with "frozen efficiency"[5] is 59% lower. Before any technological wizardry, just discerning valuable ends – that is to say, by applying the understanding that energy planning must be a "normative approach that defines what is desirable" – meant saving nearly two-third of the projected demand by the year 2000. Incorporating the fruits of the end-use analysis, namely the five efficiency improvements and energy substitution measures noted above accounted for the remaining 41%.

In sum, as compared to the LRPPP projection of 47.5 TWh the DEFENDUS scenario shaved off 12.6 TWh with clearly defined ends to guide the planning process; 8.6 TWh through energy efficiency and carrier substitution; and a further 8.3 TWh by computational changes to arrive at the correct base year demand level. Overall, DEFENDUS scenario charted out a way to require only 38% of the electricity projected by the LRPPP model, while crucially, accomplishing the development, equity and sustainability goals that remained unmet under the LRPPP projections.

6.3.2 Political Economy of the DEFENDUS Framework

The DEFENDUS framework proposes fundamental changes to the political economy of economic development and energy policy. Its basic insight for energy planning is that generation and consumption of energy are not the desired ends. Instead they are means whose desirability is contingent on their usefulness for ends found to be valuable by the community. This recognition of ends as prior to means with regard to economic development and energy planning bolsters the view that maximizing energy generation is not a prerequisite to socially valuable ends.

This kind of thinking points to a dramatic realignment of roles and relationships among entities involved in energy planning and policy. It also enriches the vocabulary

---

[5] A scenario where the demand for the year 2000 is arrived at by NOT considering any of the efficiency or substitution measures. The only input from the LRPPP, are the clearly defined ends for the future discerned from the development focus.

accorded legitimacy in such matters. Whereas the conventional arrangement tended to involve only technocrats and the political elite, and to consider and debate numerical goals, the DEFENDUS framework enriches the vista by insisting on normative reflection on ends heard in the voices and deliberations of both the lay and expert. As seen in the above illustration of the DEFENDUS framework, more equitable and sustainable energy-society relationships can be built by complementing normative reflection on ends with sound technological and financial means devised with those ends in mind. In doing so DEFENDUS challenges the validity of enclosed institutional arrangements for decision-making, such as for instance, the Planning Commission of India, which has been characterized as a venue for "extracting the politics out of political decision-making" (Chatterjee 1993).

6.3.3 Limitations of the DEFENDUS Framework

In the context of the late 1980s, responding to the dominant needs of society, the framework as applied in Karnataka emphasized domestic lighting, employment generation and irrigation. While similar "needs" exist and remain important as the focus of economic organization, there is also a considerable and growing proportion of the population now that lies beyond such "basic" needs. A question that follows is how the energy-development relationship as it pertains to this growing population is to be integrated into energy planning? Surely, the assumption cannot be that this growing population of affluent India (China or elsewhere for that matter) remains on a growth path seeking to emulate energy and material throughput of their peers living in the OECD countries. Even the most efficient society, Japan, remains an ecological debtor living beyond its ecological means. Thus, we propose that the framework's "development-focus" can be taken further.

Indeed, Reddy (1990: 49, italics in original, underline added) himself, two decades ago previewed this question, but perhaps finding it less pressing in the India of that time, thought it sufficient to assert the power of higher authority to make his case:

> If opportunities for efficiency improvements are systematically identified and exploited wherever cost-effective, the magnitude of energy demand can come down sharply. In this context, energy supplies need not become a constraint on growth. If fact, a thought experiment shows that if the most energy-efficient technologies that are either commercial today or near commercialization are deployed for all activities, then we can achieve a level of energy services or activities corresponding to Western Europe in the 1970s with only a slight increase in their per capita energy requirement. Hence, it is not the magnitude of energy that is a constraint on the achievement of significantly higher standards of living. *Of course, this process cannot go on indefinitely. Ultimately, we must accept what Mahatma Gandhi said: "The world has enough for every man's need, but not enough for everyone's greed!"*

The first reaction to this statement may be to wonder whether the idealization of Western Europe, albeit in a thought experiment, is not perpetuating a development discourse that has devalued the experience and unique possibilities of non-European or the remaining non- Europeanized contexts (Esteva 1992). The more substantive point, however, pertains to the assertion of the Gandhian axiom. The DEFENDUS framework, it seems, after exploring and charting a very useful course, comes to rest at this point. Its treatment of the question of development (much more insightful than the conventional approaches to energy planning) confesses, plainly, that efficiency is necessary but insufficient. A sustainable world includes but lies beyond the possibilities that efficiency will allow. As Reddy (1990) offers, "ultimately we must" collectively distinguish between "need" and "greed," and commit ourselves to addressing the former. This confession does not diminish the valuable distance travelled by DEFENDUS, but instead helps point us to a path that needs further wear. It is proposed here that the capability approach can valuably extend DEFENDUS for riding on this path that lies ahead. We will revisit this matter in greater detail.

The second area in which the DEFENDUS framework can be extended relates to its use of normative evaluation to not just identify end-uses but to also examining how to energize these end-uses. As it stands, the DEFENDUS framework employs only technical insights to tease out energy efficiencies and possibilities for substitution. The proposal here is that there is benefit to applying a normative evaluation to end-use analysis. Consider the case of domestic or commercial lighting. While end-use analysis as presently available in the framework readily reveals the possibility of more efficient lighting technologies, it is silent on the more normative questions of duration and/or extent of lighting. This is a space with considerable scope for mitigating ecological burdens. Perhaps, in the context of two decades ago, this might have been less pressing in energy discussions about India, but today, especially in booming urban centers, silence on this matter can be usefully ended. Put another way, does efficient lighting exhaust energy sustainability options for say, the Las Vegas strip or Times Square?

Finally, despite insightful analyses, the DEFENDUS framework has had limited success in actually influencing energy planning and policy. The balance of *power* (pun intended) still remains with the dominant discourse and its megamachine organization. In essence, while DEFENDUS brings great clarity to the alternatives, it is yet to find success in affecting the changes it wishes to see in the world. Accomplishing those ends will require its integration with efforts that have explicitly strategized to dislocate the hegemony of existing energy-development discourse and practice.

6.4 The Human Development and Capability Approach

In this section we discuss how the capability approach can be used to extend the DEFENDUS framework. We commence however, by introducing key aspects of the approach. The study of development economics emerged in the middle of the twentieth century. The subject as it emerged was marked by "an overarching preoccupation with growth of real income per head" (Dreze and Sen 2002: 34). Thus predisposed, the post-war era relied on a relevant, but nevertheless limited positive association between human well-being and per capita income as the overwhelming constraint informing public policy. In short, growth in income per head tended to be institutionalized as an end in itself. The capability approach has emerged as an effort to expand Development Economics (DE) beyond this "preoccupation with growth of real income per head." Tracing the pedigree of this motivation, Dreze and Sen (2002: 34) note that

> while...Smith and Mill, did indeed write a great deal on the growth of real income per head, they saw income as one of several means to important ends, and they discussed extensively the nature of these ends – very different as they are from income.

This essential idea of the capability approach, perhaps owing to its basic common sense, has a long history (Sen 1999: 13–14). Building on the distinction between ends and means in economic matters, and extending it to recognize the importance of intrinsic diversity among individuals, their societies and normative motivations for action, the capability approach offers a wider, informationally richer basis for evaluation and prescription of public policy than the dominant growth centered world view.

Its key contribution in this regard is the notion of *Development as Freedom* (Sen 1999). Where human development is conceived as the enhancement of the "freedoms," also referred to as "capabilities," enjoyed by people "to do" and "to be" as they find valuable. Thus, within this framework, the evaluation of the outcome of economic development efforts or prescription of policies for the same, is to turn on the expansion of valuable human freedoms, and not simply the expansion of per capita income or commodity ownership, or resource use – all of which are at best means to more substantive ends (Sen 1987, 1992, 1999). We can already see that this perspective has parallels with the "end-use" orientation embodied by DEFENDUS in the context of energy planning. But, before we build on this complementarity in later sections, let us briefly define key ideas and terms used to discuss the capability approach.

6.4.1 Functionings, Capabilities and Agency

The conceptual building blocks of the capability approach are comprised of three main themes, *viz.* "functionings," "capabilities," and "agency" as they relate to the realization of human well-being (Sen 1992). To start with then, well-being is defined simply as the "quality (well-ness, as it were) of the person's being." Functionings are defined as the interrelated "beings" and "doings" that comprise a person's life – her being. It is perhaps impossible to arrive at an exhaustive list of functionings because every individual circumstance varies from the next (Sen 1992). Nevertheless, an illustration of elementary functionings includes being well nourished, doing one's livelihood, being healthy, among others, while more complicated illustrations include functionings such as having self-respect or being happy. "The claim is that functionings are constitutive of a person's being, and an evaluation of a person's well-being has to take the form of an assessment of these constitutive elements" (Sen 1992: 39).

The notion of capabilities is closely related to functionings. The term itself can be seen to have come from the usage: "*capability* to function" (Sen 1992: 40, italics in original). The term "capabilities" refers to the universal set of functionings from which a person is free to choose the functionings that are ultimately materialized. Perhaps an analogy will be helpful here. "Just as the so-called "budget set" in the commodity space represents a person's freedom to buy commodity bundles, the "capability set" in the functioning space reflects the person's freedom to choose from possible livings" (Sen 1992: 40). One could also, in a similar vein, think of the capabilities as the cloud of all possible functionings. In that sense, it is the repository of "freedom" – thus *Development as Freedom* – enjoyed by the individual from which to choose their valuable doings or beings. The expansion of capabilities or freedoms in the capability approach serves two crucial functions. One, the expansion of freedom is "constitutive" of development. In this sense, desirable social arrangements are those that enhance freedoms enjoyed by individuals who live in them. Second, capabilities are instrumental to the process of development. In this sense, the expansion of freedoms enjoyed by individuals in addition to being an end in itself, is also crucial to facilitate the process of development (Sen 1999).

The third central concept in the capability approach is "agency" of the individual. An "agent" is defined "as someone who acts and brings about change, and whose achievements can be judged in terms of her own values and objectives, whether or not we assess them in terms of some external criteria as well" (Sen 1999: 19). The concept of agency is used within the capability approach to reflect the individual's ability to pursue ends that "normally" include the person's own well-being but can go beyond them as well. As Sen (1992: 56) notes "a person's agency achievement refers to the realization of goals and values she has reasons to pursue, whether or not they are connected with her own well-being." This actually is an important understanding of agency, perhaps building on Sen's earlier (1977) critique of the behavioral foundations of modern economics that presume individuals to be self-interested, utility-maximizing automatons.

6.4.2 *The Capability Approach and the Environmental Crisis*

The capability approach has elicited interest from the research, advocacy and policy community in various fields, including health, education, technology studies and human rights, among others. This interest extends to investigations seeking responses to environmental degradation – the close and infamous companion of the modern economic development experience. One application of the capability approach in this area emphasizes the importance of nature as a *source* of freedoms that are constitutive of human wellbeing (for example, Scholtes 2004). This line of work informs environmental policy through a systematic exploration of the linkages between ecosystem services and capabilities and functionings (for example, Duraiappah 2004;

Millennium Ecosystem Assessment 2005). The underlying logic here is that environmental degradation diminishes or degrades ecosystem services and therefore undercuts capabilities and functionings valorized by individuals and communities. Ecological health in essence enables valuable freedoms and its degradation curtails such freedoms.

Another application of the capability approach addressing the environmental crisis takes a "step back and explores the application of the capability approach to address the systemic causes of environmental deterioration" (Mathai 2004: 2). The "systemic causes" in this context, refers to the discourse of Cornucopianism and the capability approach is seen as diluting the force of this modern imperative. If an analogy is helpful, this proposal turns on the note that the capability approach is to economic development what end-use thinking is to energy planning. Just as the focus on end-use curbs the growth fixation of energy planning as demonstrated by the DEFENDUS illustration, we propose that a focus on capabilities and functionings has the ability to temper the growth fixation of the economic system.

By placing the onus on *ends* that are actually realized (functionings) or the capability to choose from a set of valuable functionings the capability approach can, in collaboration with a framework like DEFENDUS, make creative contributions:

> Focusing on human freedoms contrasts with narrower views of development, such as identifying development with the growth of gross national product, or with the rise of personal incomes, or with industrialization or with technological advance or with social modernization. (Sen 1999:3)

This tempering of the exclusive fixation on economic growth or its other surrogates eases their non-negotiability and stranglehold on economic development and energy policy:

> the rambling development creed impedes any serious public debate on the moderation of growth. Under its shadow, any society that decides, at least in some areas, not to go beyond certain levels of commodity-intensity, technical performance or speed appears to be backward. (Sachs 1999: 42)

In this context the capability approach, just as the DEFENDUS, can be a path with creative possibilities for responding to one of the fundamental drivers of the environmental crisis.

6.4.3 Similarities Between the Capability Approach and DEFENDUS

The capability approach and DEFENDUS emerged to counter, respectively, the development and energy policy status quo. Both approaches give precedence to a notion of end-use, which is developed to different extents and using different vocabularies in each case. Between the two, the capability approach has offered a more comprehensive discussion and elaboration of the idea. Its language of "capabilities" and "functionings," can improve, we offer, the gains derived from the extant association of end-use and energy planning.

A second related parallel is that both frameworks explicitly recognize deliberations on "end-use" to be normative exercises. The DEFENDUS framework recognizes the "goal-oriented" nature of systems involving human beings and the instrumental role of energy planning to help the human system toward its goals. It notes that "every goal implies, choices, values and preferences" and underscore the necessity of democratic practice for negotiating the same (Reddy et al. 1995a: 17). The capability approach does the same and more. The capability approach's articulation of *"Development as Freedom"* is crucial in this respect. First, freedom and its expansion are *constitutive* of development; they are its "primary ends" (Sen 1999: 36). As we will soon see, freedoms have great instrumental value.[6] But its constitutive role takes precedence to that instrumental value. The resulting shift in evaluative emphasis is crucial. For instance, it's the difference accorded to starving and fasting. The moral outrage elicited by the former and the moral stature accorded to the latter, ride crucially on the subjects' *freedom* "to do" and "to be" as s/he values.

Finally, both frameworks reason that normative information on ends can only be garnered through involvement of the complete human personality. The instrumental value of freedom is crucial here. Consider as an illustration the carte blanch pursuit of functionings. It is apparent that the confounding array of possibilities will make it impractical. Thus, prescriptive use of the capability approach requires "selection and weighting" (Sen 1992: 44) of the capabilities *and* functionings to be pursued. Institutional spaces and practical access to political freedom are in turn required to facilitate this process of selection and weighting.

6.4.4 The Capability Approach's Contribution to Extending DEFENDUS

While the DEFENDUS strategies of energy-services and energy-efficiency carry us a valuable distance in the quest to reduce energy demand, they can only go so far. They have to be complemented with the ability to distinguish between "need" and "greed" if social and ecological viability is valued (Reddy 1990). So the question to be answered is what criteria are available to distinguish quantity and quality of energy services demanded as "need" or "greed." The answer we propose here has two parts. The first part proposes an information space offering greater resolution to guide normative decision-making. In this regard the vocabulary of capabilities and functionings is more helpful than a simple "end-use" orientation. Second, the crucial question of institutional structure in the framing of economic development and energy policy must also be addressed.

---

[6] In addition to "political freedoms" Sen (1999: 38) highlights "economic facilities;" "social opportunities," "transparency guarantees" and "protective security" as instrumentally important freedoms.

As identified above the DEFENDUS framework asserts Gandhi's axiom of distinguishing between "need" and "greed." However, it does not offer criteria to help the policy process differentiate energy demand along these lines. Thus, *any* level of demand, in the logic of the DEFENDUS framework is reasonable, so long as it fulfills *an* end-use or provides *an* energy service. One might usefully counter this allegation with the argument that the "DEvelopment Focus" of DEFENDUS serves the function of scrutinizing energy demand and distinguishing it into "need" or "greed." Indeed the DEFENDUS calculations for the illustration discussed above applied a development focus to distinguish between "need" and "greed." Thus, it might be the case that the practical gains to be derived by informing the DEFENDUS with "capabilities" and "functioning" are limited. While this argument has merit, we hasten to suggest that the "development focus" may be limited because Development is organized as an insatiable project. It is not organized as a pursuit of basic needs or elementary capabilities. Such goals, while furnishing the rhetoric used by politicians, are incidental to the insatiability of Cornucopianism and the megamachine organization. The development project is "like the demon *Bakasura* who had an insatiable appetite; no matter how much he was fed, he always wanted more" (Reddy 1990: 12).

In this context the capability approach's engagement with the notion of end-use can usefully enrich the DEFENDUS framework. When end-use energy services are discussed as valuable capabilities and functionings, the quantity and quality of energy services demanded have additional criteria by which to be judged, individually and socially. They need no longer be insatiable like *Bakasura's* appetite. Borrowing useful vocabulary from Galbraith's "affluent society," economic development and energy policy planners have to ask whether the "preoccupation with productivity and production" enhances *valuable* capabilities. "*More* elegant cars, *more* exotic food, *more* erotic clothing, *more* elaborate entertainment" can all without quarrel be end-uses, but "more" is an insufficient basis for public policy from the vantage of furthering valuable "beings" and "doings:"

> The general format of "doings" and "beings" permits additional "achievements" to be defined and included. [Thus] there is no escape from the problem of evaluation in selecting a class of functionings – and in the corresponding description of capabilities. The focus has to be related to the underlying concerns and values, in terms of which some definable functionings may be important and others quite trivial and negligible. (Sen 1992: 44)

The need for such "evaluation and selection" (Sen 1992: 44) makes explicit the criteria for valuation, which could without concern, remain hidden and unquestioned when deliberation turns on only "end-use."

The resulting difference in the quality and content of valuation can be emphasized with an illustration. Reviewing the various lighting and display arrangements of rapidly proliferating "shopping malls" in Bangalore or New Delhi the end-use energy service perspective would focus on the scope for more efficient lighting, motors or display devices and perhaps even the possibility to source some of that enormous appetite for electricity from renewable energy. This approach, as we have seen with the DEFENDUS illustration earlier, is valuable. But it is also limited. An evaluation of these lighting and display arrangements from the perspective of their ability to enhance valuable freedoms takes the deliberation to a more foundational level. Can such energy intensive and centralized arrangements for shopping expand valuable capabilities and functionings of residents in Bangalore or Delhi? Are "shopping malls" in keeping with "underlying concerns and values" in a context of chronic electricity shortages and blackouts, poor urban air quality, traffic congestion, vulgar levels of social inequality and rampant displacement for infrastructure projects? This expanded evaluation of the energy intensity and proliferation of shopping malls from the vantage of the capability approach could beneficially extend the DEFENDUS framework.

Thus, Development *as freedom*, as opposed to "Development as *possibly* insatiable end-use energy service demand," brings space for introspection that could be of value for policies to address an environmental crisis birthed by insatiable appetites.

6.5 Extending the Capability Approach: The Sustainable Energy Utility (SEU) as a Template

Thus far our discussion on complementing DEFENDUS with the capability approach has remained silent on a crucial criticism on the latter. In this section we outline the criticism, in the context of this chapter, and present some ideas to address it.

Our proposal thus far has emphasized the role of capabilities and functionings to address the insatiability of economic development, which derives primarily form the latter's roots in utilitarian welfare analysis. However, similar to utilitarianism the capability approach comes from the liberal tradition of giving precedence to the individual (Taylor 1995; Stewart and Deneulin 2002). The resulting problem, in the context of the proposal being made here, manifests as a tension between *individual* valuations of capabilities and functionings and the need for *social* responses to the environmental crisis. Can valuations of capabilities and functionings among individuals and the agency vested in them foster a *collective* outcome of sustainability? While it is indeed possible that individual valuations and agency conducive to sustainability might emerge, it is equally possible that the opposite of these will emerge as well. And as a collective tendency the latter are likely to have the upper hand given that they are aligned with the existing discourse of Cornucopianism and the megamachine organization. Thus, a question in need of more clarity is how does social structure interact with the selection and weighting of capabilities and the scope of individual agency?

For its part the capability approach acknowledges a close relationship between social structure and the individual. It holds that "the freedom of agency that we individually have is inescapably qualified and constrained by the social, political and economic opportunities that are available to us. There is a deep complementarity between individual agency and social arrangements" (Sen 1999: xi–xii). The capability approach lays particular emphasis on political freedoms. Given that individuals are diverse and prioritize capabilities and functionings differently "there is room for explicit valuation in determining the relative weights of different types of freedoms in assessing individual advantages and social progress" (Sen 1999: 30). "Indeed, one of the strongest arguments in favor of political freedom lies precisely in the opportunity it gives citizens to discuss and debate – and to participate in the selection of – values in the choice of priorities" (Sen 1999: 30). Crucially, social arrangements are also recognized as shaping individuals' value formation and thereby the direction of their agency. The "individual conceptions of justice and propriety, which influence the specific uses that individuals make of their freedoms, depend on social associations—particularly on the interactive formation of public perceptions and on collaborative comprehension of problems and remedies" (Sen 1999: 31). Thus the capability approach offers a wide set of circumstances and ways in which society and individuals interact in the formation, weighing and selection of values.

While recognizing the precedence of social structure in these ways, the capability approach retreats to *individual* agency as the prime mover and the *individual* as the unit of moral concern. After highlighting the importance of collective life it retreats to the individual as the locus of its attention. It does not appear to accord normative content and agency to social structure (Deneulin 2008). The problem this poses for the proposal we make in this chapter is the follows: While the capability approach acknowledges, "that individual conceptions…depend on social associations," it does not offer criteria by which to evaluate these "social associations" aside from overarching endorsements of democracy and expansion of individual freedom. Nor does it offer the means to guide the negotiation and transformation of social associations.

How is a "collaborative comprehension of problems and remedies" regarding sustainability and equity to be fostered and actions sustained in the face of the overwhelming power and influence of the conventional development discourse and its institutions? Effectively challenging this status quo requires more than an endorsement of democracy and individual freedoms. The normative content and agency of such social structures must also be recognized, evaluated and used to inform alternative structures that are more sustainable.

Incorporating this recognition of social structure is crucial to our proposal of applying the capability approach to extend DEFENDUS. The informational richness that the capability approach brings to extend the end-use analysis in DEFENDUS and overcome the possibility of an insatiable demand for end-use energy services is possible only within a language that valorizes sustainability. Without the benefit of a sustainable and equitable social structure as a precondition for evaluation and selection of valuable beings and doings, an engagement with and deliberation of values may well be beneficial in some isolated, individual ways, but is unlikely to be an adequate collective response to the environmental crisis. In short, there is a need for an explicitly *sustainable "structure of living together"*[7] to respond to the environmental crisis.

---

[7] The term was used by philosopher Paul Ricoeur while referring to the notion of institutions: "By institutions, we understand the structure of living together as this belongs to a historical community, a structure irreducible to interpersonal relations and yet bound up with these" (as quoted in Deneulin 2008: 111).

6.5.1 The SEU as a Template for a Sustainable Structure of Living Together

The DEFENDUS and capability approach frameworks helpfully move us toward a human-centered energy, environment and development discourse. However, while proposing valuable alternatives they remain largely silent on the practical task of dislodging the hegemony in energy-society relations of extant Cornucopianism and the megamachine organization of society. As discussed in the previous section, we know that confronting power in this manner requires sustainable *structures of living together*. The challenge remains imagining and building such institutions. The SEU offers an innovative policy template for designing and building sustainable structures of living together. The first officially designated "Sustainable Energy Utility" was legislated into law in the U.S. state of Delaware in 2007[8]:

> The most important feature of the SEU is that energy users can build a relationship with a single organization whose direct interest is to help residents and businesses *use less energy* and *generate their own energy cleanly*. Directly put, the SEU becomes the point-of-contact for efficiency and self-generation in the same way that conventional utilities are the point-of-contact for energy supply. (SEU 2007: 2, italics in original)

In contrast to centralized and socially isolating structures for living together organizing the energy-society relationship, the SEU seeks an energy system that is organically embedded in society and its normative deliberations on desired ends. Not only does the SEU template realign energy-society relations, it crucially offers practical proposals to resist the conventional development discourse and its institutional momentum. In this effort, it enlists the concepts of "commonwealth economy" and "community trust," to rejuvenate commons' institutions, norms and practices and to transform the political economy of power in the interest of sustainability and justice (Byrne et al. 2009).

6.6 A Commonwealth Economy

A critical challenge facing investments in energy-efficiency, conservation or self-generation using renewable energy technologies is the initial investment required, despite "fuel" and operational expenses being "free" or negligible. This represents a barrier to participation that limits investment and control to public or private entities with deeper pockets. As a result the participation of ordinary businesses and families remains limited and the promise of these energy technologies to transform

---

[8] The Middle Class Task Force, convened by the United States Vice President Joe Biden identified the Delaware SEU as an energy policy innovation to transform the market to favor energy sustainability and employment generation. Internationally locations such as the metropolis of Seoul, South Korea have commissioned preliminary studies and the nation of Bermuda has expressed interest (Chang 2008; Podesta 2009; Byrne et al. 2008, 2009; Rahim 2010).

energy-society relations remains commensurately stunted. The idea of commonwealth economy, appropriated by the SEU template, recognizes these barriers and offers an institutional structure that channels wealth and thereby power over energy infrastructure from centers of capital to community. The operative word in the commonwealth economy is "common" suggesting a foundational commitment to a system of sharing, through which wealth accrues and is employed to invest in the infrastructure to build sustainable structures of living together.

The well-established financial logic behind the commonwealth economy is that energy-efficiency and conservation are superior financial investments. The energy they save is readily converted into future streams of income and offer favorable risk to return on investment ratios compared to stocks and U.S. treasury bonds (Ehrhardt-Martinez and Laitner 2008). Similarly, in the case of renewable energy technologies, policy innovation and social valuation of these technologies have produced many instruments[9] that valorize and monetize the benefits they offer for social and ecological sustainability. These facts undergird the successful and widespread "Energy Services Company" (or "ESCO") business model (for example see Vine 2005).

The commonwealth economy applies this logic and aggregates the financial surplus from future savings to underwrite present incremental capital investments needed to realize these future savings. The crucial contribution of the not-for profit SEU template is the institutional structure it offers for facilitating such investments and shared capital accumulation. This institutional point of contact is not mandated to generate surpluses for turning a profit. Instead it is charged to building and nurturing shared wealth. It will only succeed if its members are convinced by its promise, contribute to it and are empowered to use less and produce their own energy through its investments. In essence the SEU presents a template to shift control over shaping energy-society relations from the commodity centered cornucopian predisposition to a human-centered arrangement with shared goals of sustainability and justice.

The commonwealth economy establishes sustainability and justice as the means by which political and economic power is wrested, exercised and nurtured in the energy sector (Byrne et al. 2009). Such an institutional template enables a structure

---

[9] Faced with shrinking attention to energy efficiency, DSM and renewable energy after restructuring, jurisdictions across the U.S. began to innovate and explore measures to stem this loss of interest. Drawing on the legacy of social activism, scholarship and innovation, various policy tools to valorize and thus promote efficiency, conservation and renewable energy in the restructured environment, were developed and employed. They include: implementation of "systems benefit charge" for the promotion of environmental programs; the provision of interconnection with the grid and net-metering to accommodate diverse and dispersed generators and to reward them; implementation of "renewable portfolio standards" and "green pricing" to foster an overall demand for renewable electricity generation. In addition, states promoted the availability of information through customer education efforts and requirements for fuel mix disclosure. More recent innovations have included "renewable energy certificates" and even specifically targeted "*solar* renewable energy certificates" (Byrne et al. 2000).

for living together that can mitigate scale of throughput and pursue social equity in responding to the environmental crisis. It is crucial to underscore the alignment of incentives in this structure of living together. Using less to meet social objectives is both its reason for existence as well as its means to political and economic influence. It therefore can find strong and valuable synergies with contributions that DEFENDUS and capability approach have to make.

6.7 Fostering Community Trust

Community trust is essential to governing the commonwealth. Under the conventional centralized utility model there is no commonwealth and commensurately limited space for governance by the community. Individual customers remain atomistic entities served by a distant utility, on terms nominally assured by government regulators. In exchange for this anonymity customers agree to relinquish economic and political control over the energy system for the promise of abundant energy at negligible prices.

A commonwealth economy strives to change this. However, its success at regenerating and sustaining the commonwealth is contingent on its ability to gain the trust of the community of volunteer members that it serves. Such trust is crucial for two reasons. First, instead of the self-interest of a corporation, it is trust in each other and the institution that motivates members to participate, to pledge future savings to a commonwealth and to allocate it for energy efficiency, conservation and renewable energy investments sought by other members (Byrne et al. 2009).

The second, perhaps more substantive role for community trust is with regard to forming shared norms, values and practices that inform allocation of the commonwealth, and the success of such investments. This process of dialogue and deliberation through which the "meaning and practice of sustainability and equity are created and continually revised" (Byrne et al. 2009: 89) draws on the lubricant of community trust. By partaking of the commonwealth there is a shared sense of community and trust that every member will engage and reciprocate in ways that will lead to the regeneration of the commonwealth. Crucial resources to accomplish these goals – the values, norms and practices – that complement the technical apparatus of sustainability and efficiency can only be devised through a community that freely gives and receives of knowledge, ideas, criticisms and innovations for shaping energy-society relations (Byrne et al. 2009).

The valorization of using appropriately and sharing the savings and a consensus around these themes seeks to revive a personality type dismissed by modern economic and political thinking. The rational, utility maximizing stereotype has been criticized by the experiences of shared living (Byrne and Glover 2002; Ostrom et al. 2003) as well as by critiques of the behavioral foundations of modern economics, which identifies the modern stereotype as a "rational fool" (Sen 1977). The SEU offers a template for a social structure of living together where individual agency of the complete human personality can return after being sidelined by the "displaced person."

In this sense commonwealth and community trust are not innovations but a rejuvenation of modes of living together that have long sustained civilization. Indeed, it is the relatively recent institutionalized departure from this norm that is the innovation – or more appropriately – the aberration.

6.7.1 Ramifications of the SEU Template

While discussing DEFENDUS, one of our conclusions was the need for practical strategies to resist the hegemony of the conventional energy paradigm. Similarly, discussing the capability approach we recognized the need for "sustainable structures of living together." We argued that while deeper normative reflection enabled by the capability approach is helpful, it is silent on practical strategies to shift agency toward sustainability and justice. While there may well be individual valuations of capabilities and functionings conducive to sustainability (for example, X's decision to use public transport and sell her car) it is equally if not more probable that the opposite (for example, X's decision to buy a bigger and more powerful car) would also exist. And indeed the latter choice might be politically and economically privileged given the priorities of the extant structures of living together.

Within this context we arrived at the SEU as a template offering a collective capability explicitly to dislocate the hegemony of the conventional energy paradigm and foster sustainability and justice. Indeed the SEU template is predicated, through its foundational concepts of "commonwealth economy" and "community trust," on such a possibility. Notwithstanding this recognition and the importance of collective structures, individual agency still has choices to make; but the syntax offered by the template as a structural precondition for this individual agency is one that, unlike Cornucopianism and the megamachine organization, privileges ecological sustainability and equity both politically and economically.

Thus it is in alignment with X's valuable capabilities and functionings to invest in a smaller car or use public transport; and the commonwealth economy and community trust that helps X materialize her choice, are the structural preconditions through which X can choose to build a relationship of trust and reciprocity. To paraphrase Charles Taylor (1995), X's choices are individual in the proximate sense, but they are legitimized and enabled ultimately by a background of practices and understandings that valorize sustainability and equity that cannot be reduced to a set of acts, choices, or indeed other predicates of individuals like X; instead it is an emergent property of a social, political and economic community that valorizes sustainability and equity.

The SEU illustrates a template to help resist the hegemony of Cornucopianism. It demonstrates a practical public policy resistance strategy through which engaging valuable capabilities, functionings and exercise of agency can "become capable of bringing about large social change" (Mumford 1970: 424) in the interest of sustainability and equity.

6.8 Conclusion: A Crucial Role for Democratic Technics

The final question we address is what energy technology and infrastructure arrangements are conducive with such *sustainable* structures of living together? An important requirement of the alternatives discussed above is room for greater deliberation about the ends of economic activity and scope for that deliberation to interact and inform energy-society relations. Therefore energy technology and infrastructure choices responsive to such deliberation and engagement with values become crucial. Such a technological infrastructure differs from the extant predisposition toward big techno science, concentrated capital and state and/or corporate control shaping energy-society relations.

Investigations into the politics of technology offer insights ranging from the social construction of technology (Pinch and Bijker 1987) to a high degree of technological autonomy (Winner 1977). For our purposes we adopt a middle-path and appreciate the phenomenon of "technological momentum" (Hughes 1994) a position offering that to varying degrees and at different points in their evolution, technologies embody a negotiated settlement between social construction and technological autonomy. We offer that the position of a technology on this continuum reflects the balance between "democratic" or "authoritarian" (Mumford 1963) impulses inherent to it, the stage of its development and the course it has followed.

Technologies such as civilian nuclear power embody the authoritarian impulse, in that they are steel and concrete representations of non-democratic energy-society relationship. Civilian nuclear power, to illustrate, is elite driven and constructed and entertains little room to respond to public deliberation about ends, democratic involvement or control. Infrastructures such as civilian nuclear power, large dams or mega-fossil fuel burning technologies are predisposed to being exclusive and lend themselves meagerly, if at all, to being informed by democratic deliberation, selection and weighing of values.

On the other hand energy architectures such as *user-located* renewable energy generation lean more heavily toward democratic energy-society relations. The scale and modularity available with, say, a solar home system or energy-efficiency measures can create greater scope for energy infrastructure to be responsive to normative deliberation about values and ends. Indeed the iconic call for "soft energy paths" sounded by Lovins (1977) deeply recognized this highly adaptable, highly responsive and highly socially integrated attribute of such user-located energy technologies. The reasoning was that such democratic technics would move energy-society relations to more peaceful, equitable and sustainable directions. The proximity of end-uses and end-users inherent to democratic technics allows them to internalize deliberations on the question of ends with minimal loss of valuable information. The intervening decades have however seen the fading of this promise of democratic energy technologies with renewable technologies largely mimicking the architecture of centralized energy infrastructure (Byrne et al. 2009; Glover 2006).

Nevertheless, the availability of "democratic technics" (Mumford 1963) is widespread. Recorded by the work of Anil Gupta and the Honeybee Network is the enormous creativity at the "grassroots" among "knowledge rich but economically poor" people innovating to meet energy, material and economic needs (Gupta 2009). By default and/or by design this population of the world has continued to nurture and develop democratic technics and complementary skills, attitudes and social arrangements. Grassroots innovations tend overwhelmingly to internalize energy and material efficiencies and conservation practices while pursuing valuable capabilities and functioning. For instance, consider the energy and material savings made possible by the amphibious bicycle or the manual washing machine "powered" by an exercise cycle or the dual-use motor scooter-cum-washing machine that comes to your house. These innovations and thousands more, embody energy and material savings through the principles of frugality, multi-functionality, simultaneity and diversity (Gupta 2009).

Another crucial attribute of democratic technics' compatibility with sustainable structures of living together is their ability to redistribute agency and thus dilute concentration of power (pun intended). Over 50 years ago in the context of discussing solar energy and India D.D. Kosambi recognized that "if you really mean to have socialism in any form, without the stifling effects of bureaucracy and heavy initial investment, there is no other source so efficient" as energy from the sun (Kosambi 1960). The socialism that Kosambi talked about, much like Gandhi's,[10] leaned toward distributing the *ability* to create wealth and less on the redistribution of wealth. His idea was to avoid the conflict-ridden path of redistributing surplus or control over technological infrastructures that were built by concentrated capital and political authority. A crucial lesson offered by the study of the sea of creativity at the grassroots, echoes Kosambi (1960) and Mumford (1963), and finds that social inclusion is unlikely to be advanced by the dominant models of innovation, which first creates exclusion and *after the fact* seek to redress it (Gupta 2009).

The crucial challenge for engineers in this context is to help minimize this proclivity. The UNESCO (2010) report on Engineering identifies among the four important goals for the field, the need to

> more effectively innovate and apply engineering and technology to global issues and challenges such as poverty reduction, sustainable development and climate change – and urgently develop greener engineering and lower carbon technology.

If this is headed to seriously, an important challenge for engineers is to be immersed in the countless, diverse and *inclusive* places of innovation and application of *democratic* technics around the world and *share* their specialized skills or insights for furthering such places in pursuit of valuable, sustainable and equitable ends. Appellations of "green" or "low carbon" cannot any longer be based only on

---

[10] Gandhi's unique understanding of the implications of large-scale industrialization also shaped his own interpretation of "socialism" (Koshal and Koshal 1973: 194–197). While not a fan of private property he was also disinclined to "dispossess those who have possessions" (Koshal and Koshal 1973: 196), a strategy often utilized by socialist politics. Instead, Gandhi offered equality of man lay in the "dispersal of industry" and not in post-fact conflict over control of concentrated means of production.

narrow technical specifications of the technology. Instead, they also have to be accommodative of the praxis of sustainability and equity proposed above. While the dominant structural precondition of cornucopianism and the megamachine organization is not aligned with such an outcome, the strategies of commonwealth and community trust along with DEFENDUS and capability approach, present a synergy and *wherewithal* to build such interactions for sustainability and equity.

Acknowledgments This chapter was written while I was affiliated with the University of Delaware Center for Energy and Environmental Policy. My deep gratitude to John Byrne whose guidance pointed me to some crucial ideas used in this chapter. I also thank Cecilia Martinez, M.V. Ramana, Leigh Glover and Robert Warren for discussions and guidance on my research. Thanks to Ilse Oosterlaken and one anonymous reviewer for valuable editorial comments. This chapter expresses my personal views.

References

Alkire, S. (2002). *Valuing freedoms: Sen's capability approach and poverty reduction*. Oxford: Oxford University Press.

Bunker, S. G. (1996). Raw materials and the global economy: Oversights and distortions in industrial ecology. *Society and Natural Resources, 9*, 419–429.

Byrne, J., & Glover, L. (2002). A common future or towards a future commons: Globalization and sustainable development since UNCED. *International Review for Environmental Strategies, 3*(1), 5–25.

Byrne, J., & Toly, N. (2006). Energy as a social project: Recovering a discourse. In J. Byrne, N. Toly, & L. Glover (Eds.), *Transforming power: Energy, environment and society in conflict* (pp. 1–32). New Brunswick: Transaction Publishers.

Byrne, J., & Yun, S.-J. (1999). Efficient global warming: Contradictions in liberal democratic responses to global environmental problems. *Bulletin of Science Technology Society, 19*(6), 493–500.

Byrne, J., Bouton, D., Gregory, J., Rosales, J., Sherry, C., Boyle, T., Scattone, R., & Linn, C. (2000). *Environmental policies for a restructured electricity market: A survey of state initiatives*. Newark: Center for Energy and Environmental Policy. Prepared for the Science, Engineering and Technology Services Program.

Byrne, J., Wang, Y., Yu, J., Kumar, A., Kurdgelashvili, L., & Rickerson, W. (2008). *Sustainable energy utility design: Options for the city of Seoul*. Newark: Center for Energy and Environmental Policy. Seoul Development Institute.

Byrne, J., Martinez, C., & Ruggero, C. (2009). Relocating energy in the social commons: Ideas for a sustainable energy utility. *Bulletin of Science Technology Society, 29*(2), 81–94.

Chang, S. A. (2008, May). The rise of the energy efficiency utility. *IEEE Spectrum*. Retrieved March 7, 2010, from http://spectrum.ieee.org/green-tech/conservation/the-rise-of-the-energy-efficiency-utility/0

Chatterjee, P. (1993). *The nation and its fragments: Colonial and postcolonial histories*. Princeton: Princeton University Press.

Comim, F., Qizilbash, M., & Alkire, S. (Eds.). (2008). *The capability approach: Concepts, measures and applications*. Cambridge: Cambridge University Press.

Daly, H. E. (1990). Sustainable growth: An impossibility theorem. *Development, 3*(4), 45–47.

Daly, H. E. (1991). *Steady state economics* (2nd ed.). Washington, DC/Covelo: Island Press.

Daly, H. E. (1996). *Beyond growth: The economics of sustainable development* (1st ed.). Boston: Beacon.

Deneulin, S. (2006). *The capability approach and the praxis of development*. New York: Palgrave Macmillan.

Deneulin, S. (2008). Beyond individual freedom and agency: Structures of living together in Sen's capability approach to development. In S. Alkire, F. Comim, & M. Qizilbash (Eds.), *The capability approach: Concepts, measures and application* (pp. 105–124). Cambridge: Cambridge University Press.

Deneulin, S., & Shahani, L. (Eds.). (2009). *An introduction to the human development and capability approach: Freedom and agency*. London: Earthscan Publications Ltd.

Dinda, S. (2004). Environmental Kuznets curve hypothesis: A survey. *Ecological Economics, 49*, 431–455.

Dreze, J., & Sen, A. (2002). *India development and participation* (1st ed.). New Delhi: Oxford University Press.

Duraiappah, A. (2004). *Exploring the links: Human well-being, poverty and ecosystem services*. Nairobi: UNEP and IISD.

Ehrhardt-Martinez, K., & Laitner, J. A. (2008). *The size of the U.S. energy efficiency market: Generating a more complete picture*. Washington, DC: American Council for an Energy-Efficient Economy.

Esteva, G. (1992). Development. In W. Sachs (Ed.), *The development dictionary: A guide to knowledge as power* (pp. 6–25). Atlantic Highlands: Zed Books.

Glover, L. (2006). From love-ins to logos: Charting the demise of renewable energy as a social movement. In J. Byrne, N. Toly, & L. Glover (Eds.), *Transforming power: Energy, environment and society in conflict* (pp. 1–32). New Brunswick: Transaction Publishers.

Goldemberg, J., Johansson, T. B., Reddy, A. K. N., & Williams, R. H. (1988). *Energy for a sustainable world* (1st ed.). New Delhi: Wiley Eastern Limited.

Gupta, A. (2009). Grassroots green innovation for inclusive, sustainable development. In Augusto Lo'pez-Carlos (Ed.), *The innovation for development report: Strengthening innovation for the prosperity of nations*. New York: Palgrave.

Haberl, H., Krausman, F., & Gingrich, S. (2006). Ecological embeddedness of the economy: A socioecological perspective of humanity's economic activities 1700–2000. *Economic and Political Weekly, 41*(47), 4896–4904.

Habermas, J. (1973). *Theory and practice*. Boston: Beacon.

Herring, H., & Roy, R. (2007). Technological innovation, energy efficient design and the rebound effect. *Technovation, 27*(4), 194–203.

Hughes, T. P. (1994). Technological momentum. In M. R. Smith & L. Marx (Eds.), *Does technology drive history? The dilemma of technological determinism* (pp. 101–113). Cambridge: MIT Press.

IPCC. (2007). In B. Metz, O. R. Davidson, P. R. Bosch, R. Dave, & L. A. Meyer (Eds.), *Climate change 2007: Mitigation*. Contribution of Working group III to the Fourth Assessment Report of the Intergovernmental Panel on Climate Change. Cambridge/New York: Cambridge University Press.

Kosambi, D. D. (1960). *Atomic energy for India*. Pune: Popular Book House.

Koshal, R. K., & Koshal, M. (1973). Gandhian economic philosophy. *American Journal of Economics and Sociology, 32*(2), 191–209.

Lovins, A. B. (1977). *Soft energy paths: Toward a durable peace*. New York: HarperCollins Publishers.

Mathai, M. V. (2004, September 5–7). *Exploring freedom in a global ecology: Sen's capability approach as a response to the environment-development crisis*. Paper presented at the 4th International Conference on the Capability Approach: Enhancing Human Security, University of Pavia, Italy.

Mathai, M. V. (2010). *Beyond Prometheus and Bakasura: Elements of an alternative to nuclear power in India's response to the energy-environment crisis* (Ph.D. dissertation, University of Delaware, Newark).

Millennium Ecosystem Assessment. (2005). *Ecosystems and human well-being: Synthesis*. Washington, DC: Island Press.

Mumford, L. (1963). Authoritarian and democratic technics. *Technology and Culture, 5*(1), 1–8.

Mumford, L. (1970). *The pentagon of power: The myth of the machine*. New York: Harcourt, Brace Jovanovich.

Mumford, L. (2000). *Art and technics*. New York: Columbia University Press. (Original work published in 1952).

Munasinghe, M., & Swart, R. (2005). *Primer on climate change and sustainable development: Facts, policy analysis and applications*. Cambridge: Cambridge University Press.

Norgaard, R. B. (1994). *Development betrayed: The end of progress and a coevolutionary revisioning of the future* (1st ed.). London/New York: Routledge.

Nussbaum, M. (2000). *Women and human development: The capabilities approach* (Vol. 1). Cambridge: Cambridge University Press.

Ostrom, E. (2010). *Elinor Ostrom wins Nobel for common(s) sense*. Retrieved June 27, 2010, from http://www.yesmagazine.org/issues/america-the-remix/elinor-ostrom-wins-nobel-for-common-s-sense?b_start:int=1&-C

Ostrom, E., Deitz, T., Dolsak, N., Stern, P. C., Stonich, S., & Weber, E. U. (2003). *The drama of the commons*. Washington, DC: National Academies Press.

Pinch, T. J., & Bijker, W. E. (1987). The social construction of facts and artifacts: Or how the sociology of science and the sociology of technology might benefit each other. In W. E. Bijker, T. P. Hughes, & T. J. Pinch (Eds.), *The social construction of technological systems* (pp. 17–50). Cambridge: MIT Press.

Podesta, J. (2009). Testimony of John D. Podesta at VicePresident Biden's Middle Class Task Force. Retrieved March 7, 2010, from http://www.americanprogressaction.org/issues/2009/02/podesta_task_force.html

Rahim, S. (2010, February 11). State and local governments innovate to cut energy waste. *The New York Times*. Retrieved March 7, 2010, from http://www.nytimes.com/cwire/2010/02/11/11climatewire-state-and-local-governments-innovate-to-cut-92596.html

Reddy, A. K. N. (1990, July 27). *Development, energy and environment: Case Study of electricity planning in Karnataka*. Parisar Annual Lecture, Pune, India.

Reddy, A. K. N., D'Sa, A., Sumithra, G. D., & Balachandra, P. (1995a). Integrated energy planning: Part I. The DEFENDUS methodology. *Energy for Sustainable Development, 2*(3), 15–26.

Reddy, A. K. N., D'Sa, A., Sumithra, G. D., & Balachandra, P. (1995b). Integrated energy planning: Part II. Examples of DEFENDUS scenarios. *Energy for Sustainable Development, 2*(4), 12–26.

Sachs, W. (1992). Environment. In W. Sachs (Ed.), *The development dictionary: A guide to knowledge as power* (pp. 26–37). London/Atlantic Highlands: Zed Books.

Sachs, W. (1999). *Planet dialectics: Explorations in environment and development* (1st ed.). London: Zed Books.

Sachs, W. (2002). Ecology, justice, and the end of development. In J. Byrne, L. Glover, & C. Martinez (Eds.), *Environmental justice: Discourses in international political economy* (pp. 19–36). New Brunswick: Transaction Publishers.

Sant, G., & Dixit, S. (2000). Least cost power planning: Case study of Maharashtra state. *Energy for Sustainable Development, 4*(1), 13–28.

Scholtes, F. (2004). *Development as freedom and the protection of nature as a constitutive aim of the economy*. Presented at the 4th International Conference on the Capability Approach: Enhancing Human Security, Pavia, Italy. Retrieved February 4, 2010, from http://www-1.unipv.it/deontica/ca2004/papers/scholtes.pdf

Sen, A. K. (1977). Rational fools: A critique of the behavioral foundations of economic theory. *Philosophy and Public Affairs, 6*(4), 317–344.

Sen, A. K. (1987). *Commodities and capabilities*. New Delhi: Oxford University Press.

Sen, A. K. (1992). *Inequality reexamined*. Cambridge: Harvard University Press.

Sen, A. K. (1995a). Environmental evaluation and social choice: Contingent valuation and the market analogy. *The Japanese Economic Review, 46*(1), 23–37.

Sen, A. K. (1995b). Rationality and social choice. *The American Economic Review, 85*(1), 1–24.
Sen, A. K. (1999). *Development as freedom*. New York: Knopf Inc.
SEU. (2007). *The sustainable energy utility: A Delaware first*. Retrieved July 22, 2007, from http://www.seu-de.org/docs/final_report_4-21.pdf
Stewart, F., & Deneulin, S. (2002). Amartya Sen's contribution to development thinking. *Studies in Comparative International Development, 37*(2), 61–70.
Taylor, C. (1995). *Philosophical arguments*. Cambridge/London: Harvard University Press.
Tierney, J. (2009, April 20). Use energy, get rich and save the planet. *The New York Times*. Retrieved May 25, 2011, from http://www.nytimes.com/2009/04/21/science/earth/21tier.html
UNESCO. (2010). *Engineering: Issues challenges and opportunities for development*. Paris: UNESCO.
Vine, E. (2005). An international survey of the Energy Service Company (ESCO) industry. *Energy Policy, 33*(5), 691–704.
WCED. (1987). *Our common future*. New York: United Nations.
Wilhite, H., & Norgard, J. S. (2004). Equating efficiency with reduction: A self-deception in energy policy. *Energy and Environment, 15*(6), 991–1009.
Winner, L. (1977). *Autonomous technology: Technics-out-of-control as a theme in political thought*. Cambridge/London: The MIT Press.

Chapter 7
Marrying the Capability Approach, Appropriate Technology and STS: The Case of Podcasting Devices in Zimbabwe

Ilse Oosterlaken, David J. Grimshaw, and Pim Janssen

7.1 Introduction

The expansion of valuable, individual human capabilities is, according to the capability approach, a central aim of development interventions. Capabilities are the real opportunities or positive freedoms to achieve valuable 'functionings' or 'being and doings', examples of which are being healthy, participating in community life or travelling. One of the rationales behind this focus on expanding someone's human capabilities, is that this means empowering this person to be an agent, to be someone who is able to make choices and undertake actions in line with one's own ideals and ideas about a good human life. This may be either actions or choices that increase the persons's own well-being, or that contribute to other goals that the person finds important. Both well-being and agency are held to be centrally important in the capability approach. Because the capability approach is, says Johnstone (2007),

> essentially naturalistic and functionalist in orientation, capability analyses are able to integrate descriptive and normative dimensions in a way that is particularly appropriate to technological domains.

The capability approach provides a development perspective that allows one to quite naturally make a connection between on the one hand technology choice and the

---

I. Oosterlaken (✉)
Philosophy Section, Delft University of Technology, P.O. Box 5015, 2600 GA Delft, The Netherlands
e-mail: e.t.oosterlaken@tudelft.nl

D.J. Grimshaw
ICT for Development, Royal Holloway (University of London),
Egham Hill, England

P. Janssen
Department of Infrastructure, ARCADIS, Amsterdam, The Netherlands

details of engineering design – which may have a direct impact on what capabilities a technical artefact contributes to – and on the other hand the ultimate aims of development. The sparse body of literature that has so far made a link between the capability approach and technology is focused on ICT.[1] One explanation for this might be found in the "multi-purpose, multi-choice nature" (Kleine 2011) of ICTs, which can – at least in principle – simultaneously contribute to expanding many different capabilities and leave it up to the empowered user which 'functionings'[2] to realize. This, says Kleine (2011), makes 'ICT for Development' (ICT4D) "particularly well-suited to be a test-case for the choice paradigm in development evaluation, execution and planning." In reality, of course, tensions sometimes arise between the goals that development organizations or NGOs attempt to achieve and the choices that people make. An example is when so-called 'ICT telecenters' are being used for entertainment purposes – which could be seen as an exercise of these users' agency – while the NGOs intended the centers to be used for achieving pre-determined well-being goals, such as better health or improved livelihoods (Ratan and Bailur 2007). This raises a dilemma for the NGO to either respect people's choices and not meet their organizational goals, or to become paternalistic in its interventions. When reviewing the literature (Oosterlaken 2009), it becomes clear that challenges in or criticism of the mainstream practice of ICT4D – such as a tension between well-being and agency goals, too much emphasis on resource distribution and the dominance of an economic perspective – are amongst the reasons for authors to turn to the capability approach, in search of critical and fundamental reflection.

We believe that the capability approach does indeed have added value for ICT4D. However, in order to realize this potential it is very important to investigate the connections that may fruitfully be made with other approaches, theories and insights. As Zheng (2007) has noted, "many issues unveiled by applying the capability approach are not new to e-development." According to her a lot of existing perspectives and approaches within ICT4D, such as 'social inclusion', 'information culture' and 'information infrastructure', can "be used compatibly with the capability approach perspective of e-development." It is even desirable that such connections are made, as the capability approach

> is not a theory that can *explain* poverty, inequality or well-being [...] Applying the capability approach to issues of policy and social change will therefore often require the addition of explanatory theories. (Robeyns 2005)

What the capability approach can do is provide "a tool and a framework within which to *conceptualize* and *evaluate* these phenomena" (Robeyns 2005), it is "able to surface a set of key concerns, systematically and coherently, on an explicit

---

[1] A literature review in 2009 – for example – identified 18 publications in this area, of which 13 focused on ICT and 10 more in particular on ICT4D (Oosterlaken 2009).

[2] An important distinction in the capability approach is that between functionings and capabilities, or between "the realized [functionings] and the effectively possible [capabilities]; in other words, between achievements on the one hand [functionings], and freedoms or valuable options from which one can choose on the other [capabilities]" (Robeyns 2005).

philosophical foundation" (Zheng 2007). But "the capability approach offers little about understanding details of technology and their relationship with social processes" (Zheng 2007).

In this chapter we aim to illustrate both (a) what the added value of the capability approach for reflecting on technological development projects could be and (b) how the capability approach so applied could benefit from insights of existing theories and approaches with respect to technology. For this purpose we will use the case of the *Local Content, Local Voice* project, during which podcasting devices were introduced in Zimbabwe by the non-governmental organization Practical Action.[3] In Sect. 7.2 we will discuss what taking a capability approach towards the case would entail – to which aspects of the project would it draw attention? In Sect. 7.3 we will discuss how the capability approach relates to the ideas behind and experiences of the Appropriate Technology (AT) movement. To fully understand the complex and dynamic interaction between technology and human capabilities, so we will argue in Sect. 7.4, the capability approach should also pay attention to theories and insights from Science and Technology Studies (STS). Aspects of the case feature throughout these two sections as an illustration. In Sect. 7.5 we will discuss the topic of technology choice for our project case and we will argue that the capability approach allows us to conceptualize considerations of agency and well-being that play a role in such a choice. Let us first briefly introduce the case.

7.2 A Machine with Knowledge of Cattle Management

According to some the idea of appropriate technology "has not yet gained much ground in the area of ICT" (Van Reijswoud 2009). But our case, the *Local Content, Local Voice* project, is an example of an ICT project resulting from the appropriate technology movement. Practical Action only adopted its current name a few years ago and was formerly known as ITDG, the Intermediate Technology and Development Group. Established in 1966 by economists E.F. Schumacher (e.g. 1973) and others, this NGO has played a crucial role in the 'intermediate' or 'appropriate' technology movement that reached its peak in the 1970s and early 1980s. Since the last decade or so Practical Action is also explicitly paying attention to new and emerging technologies, such as nanotechnology and ICT. Our case is but one of Practical Action's activities in this area.

In 2007 Practical Action and its local partner organization LGDA have introduced mp3 players and podcasts in the Mbire district in the Lower Guruve area in Zimbabwe. 'Kamuchina kemombe' is the name that local people have given to these

---

[3] The main sources of information for the case study are documentation of Practical Action about the project, its predecessors and the ideas behind the project (Mika 2009; Gudza 2009; Talyarkhan et al. 2005), fieldwork for his master thesis by one of the co-authors in the period April-August 2010 (Janssen 2010) and experiences of another co-author with the case and its predecessors while working for Practical Action (reflected upon in Grimshaw and Gudza 2010; Grimshaw and Ara 2007).

mp3 players. The literal translation of this is 'a machine with knowledge of cattle management' (Grimshaw and Gudza 2010). The lessons on cattle management made available in this way have, according to project evaluation reports from Practical Action (Mika 2009; Gudza 2009), led to an increase in agricultural productivity and hence improved livelihoods for the local people. The introduction of this technical artefact took place as part of the pilot project *Local Content, Local Voice*, which builds on earlier work within Practical Action on the question of how to 'connect the first mile' – how to deal with the challenge of

> sharing information with people who have little experience of ICTs, low levels of literacy, little time or money, and highly contextualized knowledge and language requirements. (Talyarkhan et al. 2005)

Such challenges also apply to the Lower Guruve area. Literacy, for example, is 75%. In many other respects, the development challenges are big in this semi-arid area: livelihoods are mainly dependent on small-scale subsistence farming (livestock production and drought resistant crop cultivation); and the district's infrastructure services in the district are poor (no electricity, running water, telephone landline, mobile phone network or FM radio network). Traditional agricultural extension services[4] had ceased to be reliable because of poor transport and other economic reasons. One of the bottlenecks was, for example, that governmental livestock officers did not have enough time to properly train the animators interacting with the villagers. After consultation with local stakeholders the mp3 players were viewed as an additional channel for knowledge sharing rather than a replacement. Hence the process of sharing knowledge came to be regarded as "digital extension".

The *Local Content, Local Voice* project was part of a larger EC Block Grant project with the objective to improve livestock health and product value of resource poor households in the Mbire District. In the preparation phase, local people made a prioritization of possible interventions within the scope of this project. Also at later stages opportunities for participation were present. Participation was also built into the process of the sub-project *Local Content, Local Voice*. People were, for example, consulted on the proposed technical solution. This led to changes, such as the addition of loudspeakers to the device in order to enable collective listening while sitting under a tree in the village. Thus, a way of information sharing was made possible that is very much in line with local cultural practices. One of the key drivers of this approach was to minimize the impact of the technology on the power balance in the communities, in order to increase its chances of success (Grimshaw and Gudza 2010). Explicit attention was also paid to the podcasts themselves,

---

[4] According to Wikipedia "Agricultural extension was once known as the application of scientific research and new knowledge to agricultural practices through farmer education. The field of extension now encompasses a wider range of communication and learning activities organized for rural people by professionals from different disciplines, including agriculture, agricultural marketing, health, and business studies." But, says Wikipedia, "there is no widely accepted definition of agricultural extension" – the page lists 10 definitions from different sources to illustrate this. Source: http://en.wikipedia.org/wiki/Agricultural_extension, retrieved on February 11th 2011.

making sure that their contents would be understandable and relevant to local people. It was furthermore investigated what would be the best way to deal with the infrastructural challenges, for example using solar cells or batteries that would regularly need to be re-charged elsewhere.

7.3 A Capability Approach of the Case

So what would it mean to take a capability approach towards this case? Well, first and foremost it would mean recognizing that a successful development project is not a matter of merely giving access to resources like mp3 players. These are just means, and the capability approach would ask if they contribute to the expansion of valuable human capabilities. What are people now able to do and be, which they could not do and be before this project was implemented? The capability approach furthermore holds that poverty and well-being are multidimensional, an evaluation in line with the capability approach could thus take a wide range of things into account as relevant. The current project evaluation reports, however, limit themselves largely to outcomes in terms of the number of podcasts recorded and distributed, "a decrease in animal mortality", "increased milk yields from the animals" and "increased crop productivity" (Mika 2009). Of course strengthening people's livelihoods means strengthening their ability to support themselves. If successful, it would imply increasing people's basic capabilities, their "freedom to do some basic things that are necessary for survival and to escape poverty" (Robeyns 2005). As capabilities can be both an end in themselves as well as a means for the expansion of other capabilities, this may – in a positive spiral – also contribute to the expansion of further capabilities. As one farmer said: "because of my increased number of cattle and increased crop yields, I am now able to pay school fees for my children." Receiving more education may be valuable in its own right, yet it may also contribute to the expansion of more capabilities. Yet there may be other, less tangible project impacts. For example, according to another local farmer the mp3 players fostered "group work and group harmony which did not exist before. When groups ask for lessons we share experiences and ideas." So the technology seems to have improved farmers' relations with each other and with the development agents. The project also seems to have given farmers more self-esteem, as expressed by yet another farmer when asked about the project's benefits: "before the technology, if an animal was dying then I could not take action [lack of agency!], but now I can. I am happy since I am a full farmer now!" (Janssen 2010, pp. 70–71). If one attaches importance to both well-being and agency, as the capability approach does, such impacts are certainly worth taking into consideration. In short, the capability approach could provide a conceptual framework for a more comprehensive evaluation of the project.

What one would further want to look at, from the perspective of the capability approach, is the *process* that led to these development impacts. The capability approach resists viewing people living in poverty as passive patients to be helped, but rather pictures them as human agents able to shape their own lives. Hence, the literature on the capability approach pays a lot of attention to participative processes and democratic deliberation (e.g. Crocker 2008). Of course participatory methods have been part of development discourse and practice for quite some time now, and interesting parallels between this body of literature and the capability approach can be drawn (Frediani unknown date). The appropriate technology movement, in turn, has emphasized the importance of enabling people to choose a technology that will suit their needs. Practitioners like Practical Action advocate the empowerment of people to participate in the development process, so that they can choose an appropriate technology.[5] In the case of podcasting in Zimbabwe the participation of a wide range of stakeholders – including agricultural and veterinary agencies, local government, local development associations, community workers, and local village chiefs – has, according to Practical Action, been an important factor in bringing about ownership, empowerment and a high degree of uptake of the technology. Indeed it could be observed that the technology has been woven into the fabric of village life. We should note, however, that in the view of the capability approach participation is not just of instrumental importance for reaching pre-set goals such as technology adoption or even increased well-being (Frediani unknown date). Participation in collective deliberation and decision-making is first and foremost seen as being important for normative reasons; it is respectful of human agency to put people in the driver's seat of policies and projects that concern them. To determine to which degree this was realized in a case like ours, one could – for example – make usage of Crocker's classification of modes of participations (Crocker 2008, chapter 10), which this author himself applies to a case study of a small-scale development project as described by Alkire (2002). Such a detailed analysis of the degree of participation might, when applied to our case study, reveal that there is room for improvement.

Finally, the capability approach would draw attention to the differences that might exist between categories of individuals, in so far as these could influence the impact of a policy or project on the expansion of human capabilities. The valuable capabilities it proposes to promote "are sought for each and every person, not, in the first instance, for groups or families or states or other corporate bodies" (Nussbaum 2000, p. 74). However, the so-called 'conversion factors' could be such that a technology does not lead to a capability expansion for each and every individual. In our case it was acknowledged, for example, that the impact of ICT may be different for literate and illiterate people and by choosing for a voice-based rather than a text-based technology it was ensured that both groups would benefit. Another difference that may be relevant is that between males and females. This also applies to our case. For example, during group meetings men were seated close to the (mostly male) animator and the mp3 player. They were also more actively involved in the

---

[5] Academics like Chambers (1997) conceptualize this as a process of participative learning and action. His approach is grounded in many years of practical experience from which he notes, "local people have capabilities of which outsiders have been largely, or totally, unaware" (Chambers 1997, p. 131).

discussion after the broadcast. Sometimes their speaking volume was impossible to hear for the women who were sitting approximately 10m away. Furthermore, women sometimes needed to ask their husbands permission to individually go on a visit to the animator in order to listen to a podcast (Janssen 2010, p. 72). Such factors may, however subtle, influence the conversion of a technology into valuable human capabilities. From a capability perspective this may thus be worth investigating in more detail. More contextual conversion factors have received plenty of attention from the appropriate technology movement. Thus, this movement has a wealth of knowledge and experience to offer that is relevant from the perspective of the capability approach. It is to this topic that we now turn.

7.4 Appropriate Technology: Taking Conversion Factors Seriously

It is hard to accurately capture the ideas behind the heterogeneous appropriate technology movement in a few words. Nieusma (2004) summarizes it as follows:

> In part as a response to failures of technology transfer approaches, 'appropriate technologists' argued that context suitability should be central to identifying technologies relevant to poor people of the Third World and other marginalized social groups. [...] Attention to contextual particularities became one of the guiding approaches to appropriate technology and, hence, unlike technology transfer scholars, appropriate technology thinking took *design* as the point of intervention. (Nieusma 2004)

This focus on design does not mean that appropriate technology always needs to concern a tailor-made design solution. It may also mean that the design features of existing technological artefacts play a central role in technology choice for a specific context of application. In our case in Zimbabwe, for example, research was done into developing an innovative technological solution using Bluetooth technology and solar energy panels. The latter were considered because of the lack of an electricity network in the region. The Bluetooth technology would enable podcasts being exchanged between people passing each other. However, this technology gave rise to several difficulties and in the end this solution was not chosen for several reasons (which will be discussed in Sect. 7.5). The podcasting devices introduced instead by Practical Action were quite ordinary, existing devices. The important thing is, however, that this decision was only taken after different technical alternatives had been investigated in light of the context of application. There was no unreflected assumption that transferring some state-of-the art technology from the West would be the solution to the local development challenges.

Nieusma's view on appropriate technology is an example of what Willoughby (1990, 2005) calls the "general principles approach" to appropriate technology. This conceptualization of appropriate technology leads to a rather formal definition of what appropriate technology is. It merely "emphasizes the universal importance of examining the appropriateness of technology in each set of circumstances"

(Willoughby 2005). It thus stays close to the daily meaning of the adjective 'appropriate'; something – a technological artefact in this case – is always appropriate for something else. Such appropriateness may have many different dimensions, thus we should always ask 'appropriate *for what*?' A technology may be culturally appropriate, as when loud-speakers are added to enable collective listening in line with African practices. It may be appropriate for specific user groups, as when a choice is made for a voice-based technology in an area with a lot of illiteracy. It may be appropriate for an area lacking certain infrastructure, as when solar-powered devices are chosen for an area without an electricity network. And technology may be appropriate in an economic, political, ecological or other sense. The important thing to note is that the general principle approach makes no choice yet for one or the other type of appropriateness, it just claims that appropriateness is a very important consideration in all our dealings with technology.

Willoughby (2005) distinguishes this "general principles approach" from the "specific characteristics approach", which tends "to predominate within the Appropriate Technology movement itself." In this second approach appropriate technology is given a fixed and specific interpretation, for example ecologically sound, easy to use, low-cost, low-maintenance, labor-intensive, energy efficient, etc. Some of such interpretations resulted in the appropriate technology movement as a whole getting an image of being concerned only with simple, low-cost, low-tech solutions for poor countries, such as a smoke hood or gravity ropeway. In investigating if and how modern, 'high-tech' ICTs can be appropriate solutions for certain development challenges, Practical Action is clearly not sticking to this approach of appropriateness. A similar concept, namely "intermediate technology", was introduced by Schumacher and defined as "vastly superior to the primitive technology of bygone ages but at the same time much simpler, cheaper, and freer than the super-technology of the rich" (Schumacher 1999, p. 128). Schumacher (1973) put forward six criteria for determining if a technology was "intermediate". In the case of the podcasting in Zimbabwe we can say that the technology makes use of modern knowledge, was conducive to decentralization, compatible with the laws of ecology, gentle in the use of resources, and served the human person. That is, five out of six of Schumacher's criteria are met. The exception being production by the masses.[6] According to Willoughby the specific-characteristics definition of appropriateness:

> is more than a concept about the nature of technology and the way it relates to ends. It is simultaneously a normative statement (because it assumes priority for certain ends rather than others) and an empirical statement (because the practical criteria of appropriateness must be based upon some assessment of which technical means generally best serve the ends in question). Whereas the general-principles approach tends to leave the evaluation of ends and means relatively open, the specific-characteristics approach embodies the results of previous efforts to evaluate both of these factors. (Willoughby 2005)

---

[6]Grimshaw (2004) attempted to relate these criteria to the case of open source software. The main reason for this was to refute the often quoted view that "new technologies" could never be "intermediate technologies".

Of course, the capability approach would ascribe one important normative goal to technology, namely the expansion of valuable human capabilities. But it would certainly not claim that this should be the only goal. And especially if one keeps it an open question which capabilities should be promoted – as Sen does – the capability approach seems perfectly compatible with the general-principles approach to appropriate technology. It can even be argued that they share an important insight. According to the latter, one should evaluate technologies and their specific design features according to their appropriateness for the set of relevant circumstances. This is very important, as the context may vary a lot from country to country or even from region to region. The capability approach likewise emphasizes human diversity. The fact of immense human diversity is indeed one of the main reasons why the capability approach focuses on the expansion of human capabilities instead of resources as the end of development. After all, due to facts of human diversity the 'conversion factors' may be such that a certain resource or technological artefact does not lead to an expansion of the human capabilities needed to live the life one has reason to value. The appropriate technology movement, one could say, has always taken conversion factors seriously, even though its view was not expressed with the same concepts as the capability approach.

In the case of ICTs, it may even be more important than in other domains to pay attention to appropriateness. The reason is that actually two different resources are involved here: the technological artefact and the information distributed. Both of them are resources that could be inappropriate for the context of application or the envisaged users. Thus, one often faces what Oosterlaken (2009) has called a "double conversion challenge." Yet according to Talyarkhan, Grimshaw and Lowe:

> Projects connecting the first mile often assume that improved access to ICTs leads to improved access to information, which leads to improved knowledge and decision making and therefore development outcomes. Evidence from projects suggests that in many cases the information is difficult to appropriate because it is exogenous, in an inaccessible format, or not from a source people trust (Talyarkhan et al. 2005, p. 18)

In the *Local Content, Local Voice* project in Zimbabwe explicit attention was paid to this challenge (Grimshaw and Gudza 2010). The information needs of the local population were thoroughly investigated in the beginning of the project. The process by which people acquired knowledge, via agricultural extension, were mapped and key stakeholders included in all the dialogues. The podcasts were created in the local language and geared towards the least educated farmers in the community, so that the information would be understandable for everybody. When it became clear during the project that villagers sometimes still had difficulties putting the information of the podcasts to use, additional demonstration meetings were organized, showing – for example – how to treat sick cattle in the way explained by the podcasts.

According to Willoughby (1990) within the appropriate technology movement "there is a great deal of confusion about the meaning of Appropriate Technology". He sees this as one of the reasons that (p. 12) "while becoming a significant international movement Appropriate Technology has remained a minority theme within technology policy and practices." Another

> significant reason for the limits in the influence of the movement would appear to lie with the lack of a clearly articulated formal theory, the salient features of which are both universally recognized by the movement and identifiable by those outside the movement. (Willoughby 1990, p. 13)

It is to such a theoretical framework that the capability approach may be able to contribute something. It provides a general, normative view on development that is nowadays widely accepted – for example, it has been adopted by the UNDP. Moreover the capability approach, as we have argued, shares a key insight with the appropriate technology movement interpreted in Willoughby's 'general principle' sense: the importance and pervasiveness of human diversity.

7.5 Understanding Capability Expansion: STS

Not only work in the area of appropriate technology is useful if one is interested in the expansion of human capabilities by means of technology. The field of science and technology studies (known as STS) also has much to offer, namely an in-depth investigation of how technology and society *mutually* shape each other. This enables a richer understanding of the complex ways in which technologies and human capabilities are related. The understanding enabled by STS is richer in the sense that goes beyond the linear idea of a technological artefact (like a bicycle) being instrumentally important for expanding human capabilities (like the capability to move about), if only some relevant conversion factors (like being able-bodied) are met (an example mentioned by a.o. Robeyns 2005; Sen 1983). Figure 7.1 originally depicted, "a stylized non-dynamic representation" (Robeyns 2005) of how human capabilities relate to resources. But in reality technical artefacts do not only simply and straightforwardly expand the capabilities of an individual, who is free to use or not use the artefact to realize a certain functioning. It is more complex than that. A less stylized and more realistic picture of our dynamic reality would thus include many additional arrows, such as the dashed arrows added by the authors to Robeyns' scheme. One of these arrows indicates that the relevant conversion factors for certain categories of individuals could, if designers acknowledge them, influence the design of the artefact. And technologies also shape social practices and the social context at large, which in turn again can influence human capabilities and agency.

For example, power is an issue that is obviously important for anyone interested in expanding human capabilities and agency, in other words empowering people. And power is one of the issues at stake in the dynamics between technology and society. In the case of ICTs it is not only the technology per se, but also the knowledge that is communicated using the technology, which can change power relations. Danowitz et al. (1995) referred to ICT as being "loaded with an embedded virtual value system". Knowledge contains meaning which is dependent on context for its interpretation and understanding (Grimshaw et al. 1997). Implicit assumptions are

7 Marrying the Capability Approach, Appropriate Technology and STS...

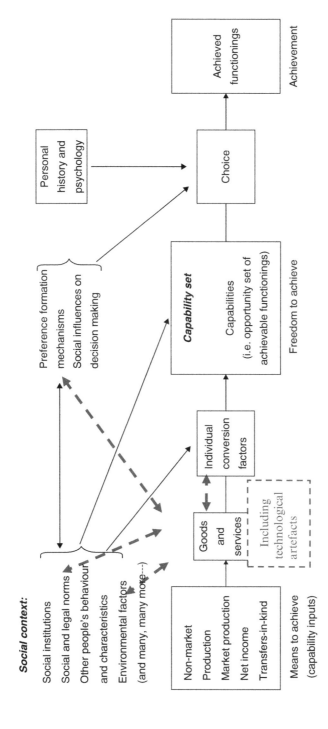

Fig. 7.1 A stylized representation of the relation between technical artefacts and human capabilities (Source: Robeyns 2005) with *dashed arrows* added by one of the authors (source: Janssen 2010)

made when that knowledge is codified and these are typically dependent on the dominant paradigm of the culture of the society where the knowledge originates. Thompson (2004) draws attention to a further dimension of the power balance with respect to ICTs; the way in which less developed countries become "locked-in" to the global networks of capital, production, trade and communications. Both media type and content source should be acknowledged as determinants of changes in the global power balance. In cases where the Internet predominates in the delivery of text based media the balance of power is away from local people. However, for technologies such as hand held voice devices which can record local content the power balance is tipped towards local people. The issue of the capability approach, power and ICT is extensively addressed by Zheng and Stahl (2011). They conclude that Critical Theory (CT), one of the streams existing within both STS and information systems research, is very useful in this respect, as it "explicitly and directly addresses the issue of technology and the distribution of power, which is exactly what is lacking in the capability approach."

Furthermore, it is important to realize that human capabilities do not only reside in human beings. This becomes clear when reading the work of Nussbaum (2000, pp. 84–85), who makes a distinction between the innate/internal capacities of a person and so-called 'combined capabilities'. The latter come about when innate/internal capacities are combined with suitable external conditions for the exercise of the functioning in question. As the capability approach is concerned with what people are *realistically* able to do and be, it takes such combined capabilities as the ends of development interventions. Similarly, Smith and Seward (2009) argue that the ontology of Sen's capability approach is "relational", as an individual's capabilities "emerge from the combination and interaction of individual-level capacities and the individual's relative position vis-à-vis social structures." However, not only individuals and social structures, but also technical artefacts are important constituents of human capabilities. The field of STS, which encompasses Actor-Network Theory (ANT), can help to gain insight in these complex and dynamic relationships between individuals, technology and social structures.[7] ANT considers *both* humans and technical artefacts to be 'actors' in a complex network that is continuously changing over time as these actors exert their influence on each other. The identities, characteristics and powers of these actors – or, what we are interested in, the capabilities of the humans in the network – depend on the precise network of relations in which they stand.[8]

In our case as well, it is not merely the podcasting device that expands human capabilities. Rather, as Janssen (2010) has described, an extensive actor network

---

[7]The example of a car can illustrate this. Basically a car remains just a specific configuration of wires, metal, nuts and bolts and so on, until it is embedded in a network with roads, gas stations, traffic rules, driving schools and the like. Only in such a network could the artefact be understood as a car, with all the powers that cars have. And only then will it be expanding people's capabilities to move about (Oosterlaken 2011).

[8]Note though that, as Elder-Vass (2008) points out, ANT generally denies the causal efficacy of social structures. Yet one could borrow some insights from ANT while still ascribing causal efficacy to three different entities: social structures, humans and technological artefacts.

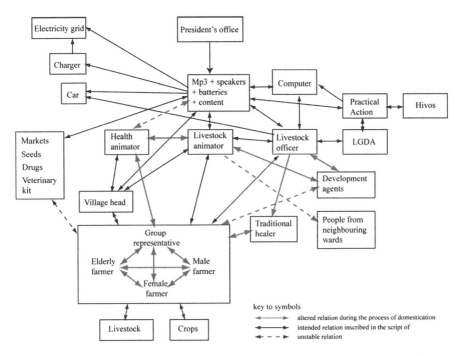

Fig. 7.2 Network surrounding the mp3 player in period April–July 2010 (source: Janssen 2010)

had to be created around these devices. The old network in which information dissemination took place was quite simple, the elements being notebooks, pencils, livestock officers, animators, community members and group representatives. That the podcasting devices were able to expand human capabilities as compared to the old situation was due to a new and more extensive network (see Fig. 7.2). This network includes – amongst others – the podcasting devices, the loudspeakers, the laptop with the database with podcast, the batteries, the charger at the head office of LGDA, the car to transport the batteries, the electricity grid available there (which is lacking, as mentioned, in the pilot area), the different government departments involved in providing the contents of the podcasts, employees of LGDA and Practical Action and the person with the right local dialect who is able to record clear podcasts. And the exact composition of and relations within the network turned out to matter for the expansion of human capabilities.[9] For example, the new and expensive cattle treatments recommended by the podcasts, such as vaccinations, were not always

---

[9] The case can also illustrate ANT's insight that technical artefacts can be seen as 'actors' in the sense that their mere presence or absence makes a difference to the course of events. For example, during the field work a health animator explained the following: "Today I gave a lesson on cholera because it was recorded in the machine" (Janssen 2010, p. 85). Thus, the mere availability of the podcast devices may 'seduce' the health animators to be guided by these, while in the absence of the artefacts they would perhaps have come to a decision to distribute a different lesson by word of mouth.

available or affordable for all the farmers and this was a limiting factor for the impact of the podcasts on human capabilities. Partly as a result of this, local people requested that indigenous knowledge be captured in podcasts as well. This was done (without verifying this knowledge in any scientific way). Furthermore, certain practices had to be developed within the network after the introduction of the devices. For example, the lessons were better understood once demonstrations accompanying the mp3 player were introduced.

As somebody's human capabilities arise in a complex interaction of this person, technical artefacts and social structures, the capability approach does not seem to support *ontological* individualism – the claim "that only individuals and their properties exist, and that all social entities and properties can be identified by reducing them to individuals and their properties" (Robeyns 2005).[10] Yet as mentioned, the capability approach *in the end* cares about the capabilities of each and every individual to lead the lives they have reason to value, not the capabilities of groups or societies at large. It thus embraces *ethical* individualism, as it makes individuals the central unit of moral concern. Because of this, a certain form of *methodological* individualism may sometimes – so Smith and Seward (2009) argue – be recommendable. The idea is that our "analysis must focus on the relative positioning of the individuals within the social structure to understand for whom different structures are differentially causal" and thus for whom – for example – certain essential capabilities may sink under some acceptable threshold level. Also in the case of technology, one should sometimes resort to such methodological individualism in order to assess technologies on their merits for different categories of individuals.

The case study can again illustrate (Janssen 2010) that it is – as ANT also acknowledges – important how a specific individual is positioned vis-à-vis the network as a whole. Obviously, people depending on livestock and crops for their livelihoods gained the most capabilities as a result of the introduction of the mp3 players. On the other hand, the basic capabilities of some traditional healers seemed to diminish due to the podcasting devices, as they more or less lost clientele to the device, people that would previously have gone to these healers with their health issues or sick cattle. The livestock animators benefited the most, since they closely related to the mp3 players and had access to its knowledge all the time. People who lived close to the animator went more often to him to demand individual lessons. Some female farmers had to ask permission of their relatives to attend group meetings or to ask the animator for individual re-playing. It is in such an analytic exercise of 'isolating' or 'highlighting' certain categories of individuals from the network, so we propose, that the specific conversion factors at work for different individuals can become clear.

Finally, STS is also useful to look at the processes of change leading to the introduction of a specific technology or a certain technological design, with certain implications for the expansion (or decrease) of human capabilities. Under the motto

---

[10] See Robeyns (2005, pp. 107–109) for an extensive discussion of different forms of individualism within the capability approach. Note though that Smith and Seward use the term 'methodological individualism' in a different way than Robeyns.

'follow the actor', ANT is interested in how different actors influence the coming about of a piece of technology or scientific insight, irrespective of conventional levels of analysis ranging from global to local, from macro to micro. Also the technology and the mostly local network in our case study has been shaped partly by at least one important actor at the macro-level, namely by the President's office of Zimbabwe. This institution supported the implementation of the ICT, but also made clear that it would hold Practical Action responsible for all disseminated content – by the way illustrating the claim made earlier that power issues matter. The original idea was to introduce a device that would allow people to record their own podcasts and disseminate them widely amongst people using Bluetooth technology. Obviously, this would make it impossible to control the dissemination of content. In the end, simple mp3 players without Bluetooth were implemented. It seems that the position of the President's office had an influence on that course of events,[11] even though technical and financial problems with respect to the original technical solution also played a role. Further research would be necessary to disentangle the factors leading up to the technology choice made.[12] What is certain is that the recording function that allowed the livestock officers to create new lessons was disabled before giving the mp3 players to the animators. The interest of the President's office has, as ANT would put it, been inscribed in the technology. This brings us to the last topic of this chapter: agency, well-being and technology choice.

7.6 Agency, Well-Being and Technology Choice

The case study of the podcasting devices introduced in Zimbabwe seems to call for some further reflection on technology choice in relation to agency and well-being. As we have mentioned, podcasts are produced and distributed on a limited number of topics only, mainly in the domains of health and cattle management. Villagers do have influence on the contents, as participatory methods were used early in the project to determine their development priorities and information needs and now that the project is running, they can make requests for new podcasting topics to the animators. Yet on the face of it, this arrangement may seem to limit the agency and capabilities of individuals in comparison with other ICT alternatives that one can imagine, of which we would like to mention two. The first was already mentioned, namely the alternative that was investigated and tested early in the project, where people would be able to directly record their own knowledge and questions, which

---

[11] See amongst others the project evaluation by Mika (2009), which recognizes the legal and regulatory environment with respect to communication technology as a factor that may work against a positive project outcome.

[12] How actors perceive a technical or financial problem, for example with either determination to tackle it or a readiness to admit defeat, may in this case have been influenced by the attitude of the powerful President's office.

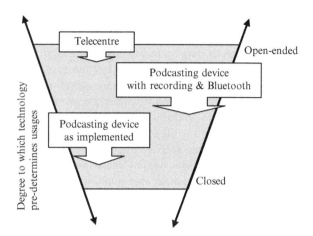

Fig. 7.3 Determinism continuum (Kleine 2011) with the discussed technological alternatives

could then – with the help of Bluetooth technology – be disseminated throughout the network of device owners. The original idea was that not only the animators, but also many of the villagers would come in the possession of a podcasting device. The second is a completely different, mainstream ICT4D alternative, namely that of the so-called 'telecenters' already mentioned in the introduction. A telecentre is basically an office with ICT equipment where people can get access to the wide variety of information offered on the internet.

We could place these three alternative technologies on a 'determinism continuum' (Fig. 7.3) as proposed by Kleine (2011), which indicates "how tightly prescribed their usage is." She rightfully notices that

> Broadly speaking, the further down on the determinism continuum a specific technology is, the more danger there is that the technology circumscribes the choices of a user-citizen more than that it widens them.

or – put differently – the higher the odds are that a technology might entail choices that do not coincide with those the individual or group of individuals would have made for themselves. This would not be judged positively from the perspective of the capability approach. Telecenters would go a long way towards the 'open-ended' extreme of the continuum and of the three alternatives mentioned, the restrained podcasting devices introduced by the project – with the disabled recording function – would be the most towards the 'closed' side of the continuum.

Kleine proposes the idea of a determinism continuum to draw attention to the importance of deconstructing the ideologies that get embedded in a technology in its context of origination. Yet if one were to use this simple and seemingly intuitive picture unwisely – with insufficient regard for the context in which the technology is to be applied – one would risk repeating exactly the mistake that the capability approach attempts to avoid, namely an excessive focus on the resources or technologies themselves, overlooking what people can actually do or be because of them. Although one could argue that *in general* more open-ended technologies are to be preferred from the perspective of the capability approach, as this in principle contributes most to expanding human agency, this may not always be the case in a concrete context of application.[13]

Recall that the question whether or not human capabilities and agency are being expanded, is within the capability approach very much a matter of 'all things considered'. In the context of the Lower Guruve area in Zimbabwe, telecenters – despite their being open-ended in principle – would in reality not contribute anything to the expansion of human capabilities of the people living there, considering conversion factors like the absence of electricity in the villages. And even when there had been electricity, it should not be overlooked that a substantial percentage of the people in this area is illiterate. Even in places where the existing 'conversion factors' are not so clearly prohibitive, there are often more subtle factors in play that still make that the telecenters do not live up to the expectations (Ratan and Bailur 2007). A nice example can be found in Rhodes (2009), who quotes the manager of an African telecenter which was supposed to be helpful to local entrepreneurs:

> We tried, and everywhere we went, at meetings and conferences people told us how good the Internet is, how we can find customers, we felt very stupid because we know people are using the Internet to help them with business, but we could not do it. We know we can do market research with the Internet, but how can we do this, we cannot understand how. (Rhodes 2009)

Here it is not so much the technical artefact, but the information itself that seems to invoke conversion problems (recall the 'double conversion challenge' mentioned in Sect. 7.4). In the *Local Content, Local Voice* project, however, great care has been taken to ensure that the podcasts are understandable to the local people and directly applicable in their everyday life. The first evaluation results can indicate that it has led to people reaping a higher income from their livestock and improved health, which could contribute in turn to expanding people's capabilities to lead the lives they have reason to value.

Something similar might be argued for podcasting devices with a recording function and Bluetooth technology; this technology is more open-ended in principle and can thus be placed more towards the desirable end of Kleine's determinism continuum. As compared to the technological alternative actually introduced in the Mbire area, it has the potential to contribute more to the agency of local people, as this alternative would allow them to record and disseminate their own knowledge and messages, without having to depend on the willingness of some employee of an NGO to grant a request to address a certain topic in a new podcast. However, *in reality* these potential agency benefits may never have been fully realized, considering

---

[13] In her article Kleine (2011) presents not only the determinism continuum, but also her wider approach to operationalizing the work of Sen. To that end, she has developed the Choice Framework, which carefully considers aspects of the context in which technologies are applied. Kleine would thus be the first to agree that her determinism continuum needs to be used without excessive focus on the technology in isolation from the context of application.

the pending pressure of the President's office to intervene with the work of the NGO and the distribution of the devices in case of unwelcome recordings of a political nature. Strategically speaking, the devices that have actually been introduced by Practical Action may not be optimal from the agency perspective, but they are arguably better from the well-being perspective, as they seem to contribute in a durable way to the enhancement of local livelihoods and thus to the expansion of a range of capabilities and so-called 'functionings'. However, Kleine rightfully notices, for more closed applications the litmus test should be whether the choices embedded in the technology align with the choices of the end users. To achieve this, Kleine argues that especially for more closed-ended technologies, there should be user participation in the decision-making process: "the more users' choices will later be locked in by the technology, the more the users' choices must already be integrated in the design process" (Kleine 2011). Participation is thus central if we want to respect people's agency, also in the process of engineering design and technological choice. Note though that this may make the 'scaling up' of a solution developed and tested in a project more problematic, as the new context of application may differ substantially from the context of origination.

Of course, the question which of the technical alternatives discussed would overall have been the best technology – taking into account practical, strategic and normative considerations – remains a question open for further debate. Around the world, ICTs have also proved their value in changing unjust and corrupt regimes, being a force of change that these regimes have found hard to control. The most important point of this short discussion on technology choice is not that a certain technology choice is definitely the best in this case. Rather, the point is that the capability approach offers a useful framework for conceptualizing and discussing such dilemmas of agency and well-being. Ratan and Bailur (2007) uncovered that such dilemmas may arise after implementation, in the usage of technology – recall their case of telecenters that are being used for entertainment purposes instead of for increasing certain forms of well-being, as the development organization intended. We have likewise revealed that such dilemmas also exist in the phase of engineering design or technology choice.

7.7 Conclusions and Recommendations

In the case studied we saw there was a certain degree of local participation in and ownership of the development process and the project went beyond making available mere resources (podcasting devices in this case). The main 'conversion factors' which could influence the development outcomes were anticipated in this project. The project has thus resulted in improved livelihoods and hence an expansion of basic capabilities of local people. Although *Local Content, Local Voice* was never explicitly conceptualized, implemented or evaluated by Practical Action in terms of the capability approach, the project thus seems – on the face of it – to be doing quite all right from this perspective. This should not surprise us, considering Zheng's observation that "many issues unveiled by applying the capability approach are not new to e-development."

A full evaluation in line with the capability approach – so we have argued (Sect. 7.3) – would take into account the multidimensionality of poverty and well-being, the degree of local participation in and control over the development process and the possible differences in development impacts between categories of individuals. Furthermore, the capability approach's concepts of agency and well-being – so we have attempted to show (Sect. 7.6) – are useful to bring out some of the issues at stake in technology choice. We also saw (Sect. 7.4) that the capability approach and the appropriate technology movement share an important insight: the importance of human diversity. What the appropriate technology movement has to offer to the capability approach is a wealth of knowledge and experience on how conversion factors can be taken into account in such a way that a technology does have the intended development impact. What the capability approach has, in turn, to offer to the appropriate technology movement is a powerful perspective on what good development is, one that has already had a widespread influence on both the theory and practice of development. In making a connection with the capability approach, the appropriate technology movement may be able to find a 'fresh' and rich conceptual framework in which to convincingly bring across its message. The capability approach can not only be enriched by the practical experiences of the appropriate technology movement, but also by the theoretical insights from science and technology studies (Sect. 7.5). Theories and approaches from this field would allow a richer understanding of how individuals, social structures and technological artefacts interact over time and co-shape human capabilities.

With the help of our case we thus hope to have illustrated both (a) what the added value of a capability approach could be and (b) how the capability approach could benefit from insights of existing theories and approaches with respect to technology. On a more practical level, the case study discussed in this chapter contains a number of lessons:

- A wide range of conversion factors may influence whether or not the introduction of a technology leads to the expansion of human capabilities. Some of these factors may be obvious, such as the absence of electricity. Other factors may be less obvious, such as women having difficulties to hear the podcasts because they are seated 'second row' during village meetings. It is best to address these factors as much as possible in the phase of engineering design or technology choice, for example by making the devices solar-powered or including a strong loudspeaker.
- It should especially be noted that ICTs often give rise to a 'double conversion challenge', as both the technology and the information to which it gives access are resources that may not always and for everybody result in an expansion of human capabilities. Attention should thus be paid not only to technology choice and engineering design, but also to the information itself. In our case, for example, the information was made available in the local language and adjusted to make it directly relevant for and applicable to the daily lives of people.

- A technological artefact or information alone does not necessarily lead to an expansion of valuable human capabilities. To achieve this, it should be embedded in an appropriate network of other artefacts and human actors. For example, podcasts on the treatment of sick cattle will not be very effective unless the recommended treatments are also made available and affordable. Also, certain (collective) practices concerning the usage of technology should develop. This may be more likely to succeed if the technology is appropriate for the local culture. In our case, relying on verbal instead of written information fitted in very well with local knowledge sharing practices.
- Open-ended ICTs in theory contribute most to expanding human agency, yet increasing well-being is also important and closed-ended technology may sometimes be very effective for this purpose. Well-being and agency should thus be explicit factors in deliberations during the phase of technology choice and design. Such evaluations should always be sensitive to the context of application and not focus too much on the technology itself. If a more closed technology is chosen, participatory processes become even more important, in order to ensure that the choices made reflect user choices closely.

Acknowledgments This research has been made possible by a grant from NWO (the Netherlands Organization for Scientific Research) and the kind collaboration of Practical Action, first and foremost in the person of Lawrence Gudza. We would also like to thank Dorothea Kleine and Sabine Roeser for their useful feedback on an earlier draft of this chapter.

References

Alkire, S. (2002). *Valuing freedoms; Sen's capability approach and poverty reduction*. Oxford: Oxford University Press.

Chambers, R. (1997). *Whose reality counts? Putting the first last*. London: Intermediate Technology Publications.

Crocker, D. A. (2008). *Ethics of global development: Agency, capability, and deliberative democracy*. Cambridge: Cambridge University Press.

Danowitz, A. K., Nassef, Y., & Goodman, S. E. (1995). Cyberspace across the Sahara: Computing in North Africa. *Communications of the ACM, 38*(12), 23–28.

Elder-Vass, D. (2008). Searching for realism, structure and agency in actor network theory. *The British Journal of Sociology, 59*(3), 455–476.

Frediani, A. A. (unknown date). *Participatory methods and the capability approach*. In Briefing notes. Human Development and Capability Association. http://www.capabilityapproach.com/pubs/Briefing_on_PM_and_CA2.pdf. Accessed 13 June 2008.

Grimshaw, D. J. (2004, June). *The intermediate technology of the information age* (New Technologies Briefing Paper No. 1). Rugby: Practical Action.

Grimshaw, D. J., & Ara, R. (2007). Local content in local voices. *ICT Update*, Issue 37.

Grimshaw, D. J., & Gudza, L. D. (2010). Local voices enhance knowledge uptake: Sharing local content in local voices. *The Electronic Journal on Information Systems in Developing Countries (EJISDC), 40*(3), 1–12.

Grimshaw, D. J., Roberts, S. A., & Mott, P. L. (1997). The role of context in decision making: Some implications for database design. *European Journal of Information Systems, 6*(2), 122–128.

Gudza, L. D. (2009). *Sharing local content in local voices; spreading the use of Podcasting - pilot project PODCASTING*. End-of-Pilot Project Report submitted to HIVOS. Harare: Practical Action.

Janssen, P. (2010). *Kamuchina Kemombe: Opening the black-box of technology within the capability approach* (Master thesis for the program 'Philosophy of Science, Technology and Society', University of Twente, Enschede).

Johnstone, J. (2007). Technology as empowerment: A capability approach to computer ethics. *Ethics and Information Technology, 9*, 73–87.

Kleine, D. (2011). The capability approach and the 'medium of choice': Steps towards conceptualising information and communication technologies for development. *Ethics and Information Technology, 13*(2), 119–130.

Mika, L. (2009). *Sharing local content in local voices; spreading the use of podcasting pilot project – Final evaluation report*. Harare: Practical Action Southern Africa.

Nieusma, D. (2004). Alternative design scholarship: Working towards appropriate design. *Design Issues, 20*(3), 13–24.

Nussbaum, M. C. (2000). *Women and human development: The capability approach*. New York: Cambridge University Press.

Oosterlaken, I. (2009). *ICT and the capability approach – A literature review and research proposal*. Paper presented at the 16th biennial conference of the Society for Philosophy and Technology (SPT 2009: Converging Technologies, Changing Societies), University of Twente, the Netherlands.

Oosterlaken, I. (2011). Inserting technology in the relational ontology of Sen's capability approach. *Journal of Human Development and Capabilities, 12*(3), 425–432.

Ratan, A. L., & Bailur, S. (2007). Welfare, agency and "ICT for Development". In *ICTD 2007 – Proceedings of the 2nd IEEE/ACM international conference on Information and Communication Technologies and Development*. Bangalore: IEEE.

Rhodes, J. (2009). Using actor-network theory to trace and ICT (telecenter) implementation trajectory in an African Women's Micro-enterprise Development Organization. *Information Technologies and International Development, 5*(3), 1–20.

Robeyns, I. (2005). The capability approach – A theoretical survey. *Journal of Human Development, 6*(1), 94–114.

Schumacher, E. F. (1973). *Small is beautiful; A study of economics as if people mattered*. London: Vintage Books.

Schumacher, E. F. (1999). *Good work*. London: Jonathan Cape.

Sen, A. (1983). Poor, relatively speaking. *Oxford Economic Papers (New Series), 35*(2), 153–169.

Smith, M. L., & Seward, C. (2009). The relational ontology of Amartya Sen's capability approach: Incorporating social and individual causes. *Journal of Human Development and Capabilities, 10*(2), 213–235.

Talyarkhan, S., Grimshaw, D. J., & Lowe, L. (2005). *Connecting the first mile; investigating best practice for ICTs and information sharing for development*. Rugby: ITDG Publishing.

Thompson, M. P. A. (2004). ICT, power, and developmental discourse: A critical analysis. *Electronic Journal of Information Systems in Developing Countries, 20*(4), 1–25.

Van Reijswoud, V. (2009). Appropriate ICT as a tool to increase effectiveness in ICT4D: Theoretical considerations and illustrating cases. *The Electronic Journal on Information Systems in Developing Countries (EJISDC), 38*(9), 1–18.

Willoughby, K. W. (1990). *Technology choice; A critique of the appropriate technology movement*. Boulder/San Francisco: Westview Press.

Willoughby, K. W. (2005). Technological semantics and technological practice: Lessons from an enigmatic episode in twentieth-century technology studies. *Knowledge, Technology, and Policy, 17*(3–4), 11–43.

Zheng, Y. (2007). *Exploring the value of the capability approach for E-development*. Paper presented at the 9th international conference on Social Implications of Computers in Developing Countries. Sao Paulo, Brazil.

Zheng, Y., & Stahl, B. C. (2011). Technology, capabilities and critical perspectives: What can critical theory contribute to Sen's capability approach? *Ethics and Information Technology, 13*(2), 69–80.

Chapter 8
From Individuality to Collectivity: The Challenges for Technology-Oriented Development Projects

Álvaro Fernández-Baldor, Andrés Hueso, and Alejandra Boni

8.1 Introduction

Throughout history, technology has been a powerful tool for development. The wheel allowed us – for example – to transport heavy loads and, more recently, mobile phones enabled us to communicate from any place in the world. Technology is also used for poverty reduction in many different ways – from water supply or electrification to developing long-distance education or telemedicine.

However, over the past decades numerous technology-oriented development projects have failed. The many agricultural modernization projects neglected and the hundreds of photovoltaic panels abandoned are just two examples of this failure.[1] In both cases technology was transferred to developing countries, but only technical issues were taken into consideration. Little attention was given to processes of technological change, thus leaving out important issues such as participation or empowerment of people. These examples show that technology, despite being important, is not the only factor that ensures the success of a technological intervention.

---

[1] Some examples can be seen in Dufumier (1996) and James (1995).

Á. Fernández-Baldor (✉)
Group of Studies on Development, International Cooperation and Ethics,
Universitat Politècnica de València, Valencia, Spain
e-mail: alferma2@upv.es

A. Hueso • A. Boni
Department of Projects Engineering, Group of Studies on Development, International Cooperation and Ethics, Universitat Politècnica de València, Valencia, Spain

It is still common today to channel such technology-oriented development aid through small-scale cooperation projects implemented in rural communities or villages. As such, development projects are not an end in themselves, but rather instruments for promoting and supporting complex processes of change and transformation. We assume that the Capability Approach provides us with the conceptual elements required to broaden our vision and enables us to go beyond specific technological results. Thus, this chapter aims to discuss technological aid projects implemented in small communities from the perspective of the capability approach.

For this purpose, firstly we examine the evolution of technology-oriented development projects in the last decades and the limitations of its current conceptualisation. Secondly, we present different aspects of the capability approach that can help us to re-conceptualise technology-oriented aid projects implemented in small communities. Special attention is given to agency and the tension between the individual and the collective. We will call our new conceptualisation of technology in the context of human development *Technologies for Freedom* (T4F). In this chapter we will present three case studies of power projects based on previous field research of the authors that illustrate the different steps of technology-oriented development projects discussed herein. Finally, we point out some characteristics that a technology-oriented development project should consider if its final purpose is to expand the freedom of community members.

8.2 History of Technology-Oriented Development Projects

8.2.1 Appropriate Technologies

India, at the end of the nineteenth century, is identified as the place where the Western concept of Appropriate Technology (AT) had its genesis. The thought of the reformers of that society was oriented to the rehabilitation of traditional technologies, used in villages, as a fighting strategy against British domination. Between 1924 and 1927, Gandhi spent a long time spreading the *Charkha*. This spinning wheel was both a tool and a symbol of the Indian independence movement. The *Charkha*, a small, portable, hand-cranked wheel, is ideal for spinning cotton and is recognized as the first equipment that was labelled as 'technologically appropriate'. Promoting self-sufficiency in cloth making represented a way of fighting against social injustice and the caste order established in India. Thus, a political conscience arose in millions of people, especially in rural areas, about the necessity of renewing the Indian native industry. This can be expressed in the famous Gandhi's words: "Production for mass, not mass production" (Kumar 1993, p. 535).

Gandhi's protection of male and female handicrafts in villages did not imply a static conservation of traditional technologies. Rather, it promoted an improvement of local techniques, the adaptation of modern technology to the environment and local conditions of India, and support of scientific and technological research oriented to identifying and solving the problems of people. The final objective was the transformation of the Indian society through an endogenous process, and not by an external imposition. Therefore, in the social doctrine of Gandhi "the concept of appropriate technology is clearly defined, even though he never used it" (Herrera 1983, p. 11).

Gandhi's ideas were also applied in China and, later, influenced a German economist – E.F. Schumacher – who introduced and popularised the Appropriate Technology (AT) term in the Western world. In order to be appropriate for developing countries, the technology would have to be small scale, simple, environmentally friendly and low cost technology. Schumacher established the Intermediate Technology Developing Group (now called Practical Action) and published amongst others *Small is beautiful: economics as if people mattered* (Schumacher 1973), which was translated into more than 15 languages.

The AT movement had a big impact. During the 1970s and 1980s many research groups proliferated in northern countries, concerned with developing and implementing technological artefacts based on the AT ideas. Although the main purpose of those groups was to reduce poverty in developing countries, they frequently dealt with issues related to the environment and alternative energy (Dagnino et al. 2006).

We can find many definitions of AT: Alternative technology, Intermediate technology, Adequate technology, Social Appropriate technology, Environmental Appropriate technology, Human technology, Help-self technology, Low income technology, etc. (Pérez-Foguet et al. 2005; Brandão 2001). Despite these different names, some common characteristics appeared in technology-oriented development projects that were inspired by the AT movement: based on community participation, low costs of services provided, small-scale, simple, labour-intensive, respectful to local culture, the environment, etc. In other words, the aim was technology that would be "able to avoid social and environmental problems derived from conventional technology transfer processes and, additionally, able to decrease the technological dependence" of developing countries on the West (Dagnino 1976, p. 86).

The preoccupation with unemployment around the world was a stimulus to the AT movement. The most significant example was the involvement of the International Labour Organization (ILO) which, at least at a discourse level, supported many case studies for evaluating the utilization and development of AT, mainly in Asia and Africa (Behari 1976; Goodman 1976; White 1974). These studies demonstrated the role that an intensive use of the local labour force could play in terms of better social and economic impact. But, at the same time, a lack of external support for the AT movement seems to be the reason beyond the scarcity of resources for researchers from developing countries to make their own contributions.

8.2.2 Critiques to Appropriate Technologies

Most of the critiques of AT are presented from the perspective that science and technology are not neutral and technology-oriented aid projects involve decisions which incorporate values. Those values can generate justice or injustice, safety or insecurity, etc., depending on the decisions made when designing and transferring technologies. Unfortunately, in most AT projects the communities are only involved in a final stage of technology transfer, as the stakeholder that benefits from technology designed in developed countries (Leach and Scoones 2006; Chambers 1997). Deepening on this disapproval, its detractors focus on the view that AT is more a movement of retired researchers in developed countries than a real initiative able to make a significant change in the South (Dagnino et al. 2006). In fact, most of the research groups in AT are located in developed countries.[2] Underlying this critique is the idea that increasing the set of technological alternatives offered to developing countries is not sufficient to change the nature of the process of creation and diffusion of technologies and knowledge. Consequently, AT will just be a downgrading of conventional Western technologies (Herrera 1983; Leach and Scoones 2006; Shiva 2009).

Furthermore, ATs are usually understood in a narrower "micro" sense, as concrete and specific technological solutions to the main problems faced by the poor in developing countries. In this sense AT has been criticized on the grounds of being excessively techno-centric and not taking into consideration problems and realities present at a macro level in developing countries (Pérez-Foguet et al. 2005).

Another critique is centred on the replication of the technology. The specific context of implementation in developing countries makes the replication of AT artefacts or models more complicated. Hence, it is difficult to transfer a technology because the context and the people who will use this technology are also different (Chambers 1997). For example, a single energy (wind, solar or hydro) technology can work perfectly in a community, but when is transferred to another community, even if it has similar characteristics (related to the wind, radiation, or water flow and jump height), it may not work. Other factors, such as power relations, social rules or gender roles play an important role in aid projects and must be considered if we wish to achieve a successful intervention.

8.2.3 UNDP Vision of Technology for Human Development

In 2001, after 11 Human Development Report (HDR) editions, the United Nations Development Programme (UNDP) published the *HDR 2001: Making new technologies work for human development*. This was the first attempt to link technology and development under the approach of Human Development. At the beginning, in the foreword, we read:

"[…] technology is used to empower people, allowing them to harness technology to expand the choices in their daily lives" and "[…] research and development addressing specific problems facing poor people – from combating disease to developing distance education – have proved time and again how technology can be not just a reward of successful development but a critical tool for achieving it". (United Nations Development Programme 2001)

---

[2]Engineers Without Borders in many developed countries; Practical Action or Tearfund in UK; Village Earth, AIDG or Whitman Direct Action in USA; Centre for Appropriate Technology in Australia; among others.

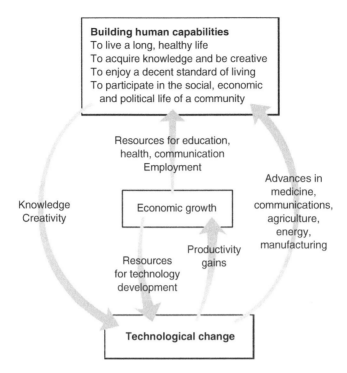

Fig. 8.1 Links between technology and human development (Source: UNDP 2001, p. 28)

Figure 8.1, created by the UNDP, shows how technological innovation affects human development in two different ways. First, directly, because products – vaccines, access to Internet, etc. – improve people's health, nutrition, knowledge and living standards, and increase people's ability to participate more actively in the social, economic and political life of a community. Thus, according to the UNDP approach one way to achieve human development is through innovations such as drought-tolerant plant varieties for farmers in uncertain climates, vaccines for infectious diseases, clean energy sources for cooking or Internet access for information and communications.

And secondly, technological innovation contributes to human development because of its impact on economic growth through the productivity gains it generates. Economic growth raises the crop yields of farmers, the output of factory workers and the efficiency of service providers and small businesses. It also creates new activities and industries – such as the information and communications technology sector – contributing to economic growth and employment creation. In addition, human development is a means for technology development, so "human development and technology advance can be mutually reinforcing, creating a virtuous circle" (United Nations Development Programme 2001).

But, as we argue in this chapter, and according to other authors (Herrera 1983; Leach and Scoones 2006; Shiva 2009), if there are no processes of alternative generation and diffusion of knowledge, the change in the lives of the poorest will not be

> **Chucura** is an indigenous Aymara community in the rural area of La Paz (Bolivia), which benefits from some privileges as it is inside a National Park. In 2003 national and international donors funded a micro hydro power plant of 21 kW, which was designed and built by a university in La Paz. The community was supposed to participate in the construction, but in the end only few families did. When the project finished, people started to benefit from all the advantages of having the plant: reduction of 50% of household expenditures for energy services, better conditions for studying, access to TV, better health due to elimination of the smoke generated by lighting, public lighting, etc. Some time afterwards, however, the plant started having maintenance problems. As nobody seemed to be responsible for maintaining it, the plant began to have more and more problems and the electricity service was interrupted very often.

Fig. 8.2 Case 1: Example of UNDP vision

sufficient. In the 2001 HDR it is not clear what "technological change" means. It is a *black box* where knowledge, creativity and economic resources are supposed to be transformed into productivity gains and advances in medicine, communications, agricultures, energy and so on. But, do these processes always involve a true community development? Is there a real change in the lives of the poorest? It seems that this 'positive' view of technology forgets concepts such as empowerment, equality, productivity and sustainability, the basic principles of human development (Ul Haq 1995). Moreover, there is no doubt that medical breakthroughs such as immunizations and antibiotics resulted in faster gains in the last century. But most of the technological innovations were produced in developed countries, and diffused and adapted to developing countries. Thus, the technological divide was maintained and the dependence increased.

Despite the fact that the UNDP has adopted, at least at a discourse level, the paradigm of human development, our main critique of the 2001 HDR is that it has not fully driven through in practice the implications of this choice. Technological innovation and economic growth are necessary, but the way in which change occurs is also important. The process of design, transfer and implementation of technology and how people participate in that process seems not to matter to the UNDP. The Chucura electrification project[3] is an example of a development project that we can consider under a UNDP approach (Fig. 8.2).

In case 1, a technology (hydro power plant) was introduced into a poor community. The energy service was supposed to improve the living standards of people. But everything appears to indicate that the project has not been successful, and it does not seem to have failed due to a technical problem. The UNDP/HDR 2001 vision assumes that a technological change (in this case a micro hydro power plant)

---

[3]The case studies discussed in this chapter are based on previous field research. For further information, see Fernández-Baldor et al. (2009) and Hueso (2007).

translates directly into human development for community members. But in development practice there are other important issues to keep in mind if we wish to achieve a sustainable and successful intervention. It is important to consider environmental and political issues, but also real participation and empowerment, as well as the inequalities and differences between human beings and the real freedoms of people to choose their well being (Dubois 2006). In Chucura no attention was paid to such issues.

We believe that the Capability Approach is an interesting framework for reconceptualising technology-oriented development projects, as it provides a more complex view on their impacts and processes.

8.3 The Capability Approach

8.3.1 *Core Concepts*

Amartya Sen's Capability Approach provides the philosophical basis of the human development paradigm (Fukuda-Parr 2003). This paradigm, promoted by the UNDP and discussed in the HDRs, covers all aspects of development – from economic growth to international trade; budget deficits or fiscal policy; savings, investment or technology; basic social services or safety nets for the poor. No aspect of the development model falls outside its scope, but the point of reference remains the widening of people's choices and the enrichment of their lives (Alkire and Deneulin 2009). This approach has four key *principles*: equality, sustainability, productivity and empowerment (see among others, Ul Haq 1995), even if other principles such as responsibility or respect for human rights also matter (Alkire and Deneulin 2009). It regards economic growth as essential, but emphasizes the need to pay attention to its distribution, analyses at length its link with the quality of human lives and questions its long-term sustainability.

The capability approach focuses on peoples' capabilities or real possibilities of leading a life which they have reasons to value (Sen 1979, 1999). *Capabilities* and *functionings* are two key concepts in the approach. Capabilities refer to different combinations of functions which can be achieved, where functionings are "the different things that a person can value doing or being" (Sen 1999, p. 3). These beings and doings together constitute what makes a life valuable. Functionings include working, resting, being literate, being healthy, being part of a community, being respected, and so forth. The distinction between functionings and capabilities achieved is a distinction between what has been realised and what is effectively possible; in other words, between achievements, on the one hand, and freedoms or valuable options from which one can choose, on the other (Robeyns 2005). Capabilities, then, are the *freedom* to enjoy valuable functions. Sen defines freedom as "the real opportunity that we have to accomplish what we value" (Sen 1992, p. 31). Capabilities are the specific positive freedoms: the freedom to do or be what one *values*. Freedom, thus, plays a substantive role in development.

> **Charía** is another Aymara community in the rural area of La Paz, with a population of 80 families. In 1996, a micro hydro power plant was built with strong participation of the community. Decisions were made in deliberative and democratic community assemblies, taking into account the views of all members. Nowadays, the plant is still working and has provided many improvements for the well being of the community. Examples are the improvement of the conditions and means for studying, the reduction of indoor smokes (caused by lighting) or the improvement in health care services. In addition, the community created a fund that guarantees the sustainability of the project. All this results in an expansion of the capabilities of the members of the community (the capability to be educated or to have good health).

Fig. 8.3 Case 2: Example of expansion of individual capabilities

In this regard, Sen's capability approach has gone beyond the utilitarian perspective when it comes to judging the way a society functions and develops (Sen 1979, 1989). Freedom is an end in itself, and not only a means for other types of uses. Therefore, for a society to develop, the main sources of freedom deprivation must be eliminated. Under this approach, the main purpose of development is to expand people's choices or, in other words, to create an enabling environment for people to enjoy long, healthy and creative lives.

One interesting thing of the capability approach is that it focuses primarily on the *process* instead of stressing the results and products of the interventions (Crocker 2008). The UNDP's argues that to be empowered, people need to fully participate in the decisions and processes that shape their lives (United Nations Development Programme 2005). According to Alkire (2005, p. 227):

> quite a few studies indicate that durable poverty reduction or enduring social change occurs when some poor persons, as well as others in their society, participate actively in development processes. Such is the strength of this finding that it has become a truism to advocate the 'participation' and 'empowerment' of persons in many dimensions.

In that 'participatory' process, the capability approach takes into account social inequalities generated by diversity (Watts and Bridges 2006), where equality does not mean equal income, but equal human capabilities (Walker 2006). We argue that also in technology-oriented development projects implemented in small communities, an inclusive and fair participation process is needed. Such a process will allow to include different points of view to count in choosing which goals to pursue.

Thus, although the UNDP has adopted the capability approach, at least at a discourse level, in practice it has not fully exploited the potential of the approach. Applying the capability approach has more implications, such as the focus of attention on participation, distribution or inequalities. At this point we can introduce a second case study, Charía, which deviates substantially from the aforementioned case. In this second step (Case 2), other issues such as public spaces of deliberation and effective participation were added to the UNDP vision (Case 1), thus ensuring the sustainability of the project (Fig. 8.3).

The project has thus been successful, but would it be possible to think of a project able to promote a deeper change? We believe so. Technology-oriented development projects should go further if they want to make communities more capable of shaping their future. A less explored concept of the capability approach, 'agency', can help us to think about a deeper change.

8.3.2 Agency

Another central concept in the capability approach is agency, defined as the ability to act according to what one values or –in Sen's words – "what a person is free to do and achieve in pursuit of whatever goals or values he or she regards as important" (Sen 1985, p. 203). An agent is "someone who acts and brings about change, and whose achievements can be judged in terms of her own values and objectives" (Sen 1999, p. 18).

Sen argues that freedom has two different aspects: opportunity and process. The opportunity aspect pays attention "to the ability of a person to achieve those things that she has reasons to value", and the process aspect focuses attention on "the freedom involved in the process itself" (Sen 2002, p. 10). The notion of capability refers to the opportunity aspect of freedom, while the notion of agency refers to the personal process of freedom (Des Gasper 2007). Freedom and agency are mutually interconnected: "wider freedoms allow agents to act and achieve the goals they value, while the exercise of agency leads to a further widening of freedoms" (Ibrahim 2006, p. 400).

As Sen's states, "greater freedom enhances the ability of people to help themselves, and also to influence the world, and these matters are central to the process of development" (1999, p. 18). So, from this perspective, development relies on people's freedom to make decisions and advance key objectives. People themselves decide upon what kind of development they would like for themselves, so that "people who enjoy high levels of agency are engaged in actions that are congruent with their values" (Alkire 2008).

Agency differs from well being in the sense that agency is not only concerned with the goals that lead to a person's own wellness or personal welfare, but to all the goals he/she has in mind (Crocker 2008). This distinction is important as one can pursue objectives that may reduce one's welfare, for example, when parents starve in order to give their children enough food. Thus, agency goals may incorporate commitments to other individuals or to causes, and occasionally their pursuit may result in actions deleterious to the individual's own well being (Sen 1985).

In this regard, Crocker makes it clear that agency "provides conceptual space for a conception of freedom and responsibility that breaks decisively with any egoism that claims that humans are no more than strict maximizers of a narrowly defined self-interest" (Crocker 2008, p. 5).

8.3.3 From an Individual to a Collective Approach

When people and social groups are recognized as agents, they can define their priorities and choose the best means to achieve them. However, the capability approach is often criticized for its excessively individualistic vision, as it is mainly concerned with 'individual capabilities' (Ballet et al. 2007; Stewart 2005; Deneulin and Stewart 2001). But individuals cannot be considered independently of their relationships with other people and with the institutions (Ballet et al. 2007). Sen (1982) also recognizes the importance of social values in affecting the acts of individuals. Furthermore, some scholars argue that belonging to a community and participating in its day-to-day life has an impact on the level of well being of the individuals (Nussbaum 1987, 2000; Alkire 2002; Robeyns 2003). This tension between the individual and the collective is particularly relevant to our work, because most technology-oriented aid projects are implemented in local communities (small groups of people living in a common location), and require collective action of their members.

Disagreements and inconsistencies seem to appear once the space for evaluation between individuals and collectives is defined. Some would argue that the capability approach takes into consideration structural processes, but the evaluations should be concerned in the realm of individuals. Collectives would be omitting the voice of the oppressed, therefore reinforcing unequal power relations. Meanwhile, critics within the capability approach would argue that one needs to engage in the understanding of collective (or social) capabilities (see Evans 2002; Deneulin and Stewart 2002; Ibrahim 2006).

But, what are collective capabilities? Stewart (2005) states that group capabilities (or collective capabilities) are just the average of individual capabilities. However, other capability approach scholars as Comim and Kuklys (2002) or Ballet et al. (2007) view collective capabilities as more than a sum of individual capabilities, as a result of social interactions. We can find a clear definition in Ibrahim (2006). She defines collective capabilities as "the newly generated functioning bundles a person obtains by virtue of his/her engagement in a collectivity that help her/him achieve the life he/she has reason to value" (Ibrahim 2006, p. 398). In that work, Ibrahim demonstrates how the poor can act together to expand and exercise new 'collective capabilities'.

At this point we can contextualise the concept of collective capabilities with a third case study in Comunidad Nueva Alianza (Fig. 8.4).

It seems to be clear that the inhabitants of the Nueva Alianza community are generating new individual and collective capabilities to a much larger scope than the people in Charía. Therefore, they also acquire a greater ability to help themselves and reach other goals they may find valuable (agency). But, is this agency individual or collective? Is there a 'collective agency'? The work of Ibrahim (2006) can be helpful here. She argues that human beings can bring about changes in their societies both through individual and collective actions. She introduces an interesting example: "Many women are subject to female genital mutilation (FGM). If a group

> **Comunidad Nueva Alianza** is an example of a community where a technology-oriented development project ignited a change for a whole group. Nueva Alianza is a community of 45 coffee producer families in the rural area of Guatemala. An international donor funded a micro hydro power plant requested by this community. Although the final amount of energy delivered did not match the initial forecast, this project represented a considerable advance on the development process of the inhabitants of Nueva Alianza. The power plant is working properly and has provided for many improvements in the well being of the community, such as reduction of indoor smokes, illuminated homes or prevention of unnecessary trips to the city to buy diesel. But most importantly, the project has served as a driving force for the community to become involved in other new projects, and now everyone in the community is benefiting of new improvements. The telecenter project or the pig farm are just two examples by which the community benefits of Internet access, new incomes or greater variety of foods. This was possible thanks to the participation and involvement of the whole community and the enthusiasm put into the projects. The success of the first project (electrification) led the community to believe in their ability to face new challenges. In this regard, community involvement was not reduced only to implement new projects for their own benefit, but they also participated in actions that would benefit others. For instance, they engaged in rural movements to claim peasants' rights or they started to support neighbouring communities to seek aid from different donors, stressing its commitment with other poor communities.

Fig. 8.4 Case 3: Example of expansion of collective capabilities (and collective agency)

of women decided to fight against FGM, not only would they be expanding their individual freedoms – especially if they were already victims of FGM – but also they would be promoting the 'collective freedoms' of women who might be subject to FGM in the future" (Ibrahim 2006, p. 405). She also argues that collective agency is valuable for generating new capabilities, but also intrinsically important in shaping and pursuing the individual's perception of what is good.

If we go further in the Nueva Alianza case we find evidences of the collective agency described by Ibrahim. The electrification project has served, as explained in Fig. 8.4, as a driving force for the community to become involved in other new projects. But it has also helped to claim the community's rights, to engage in rural movements and to support other communities. We can argue, based on this case, that in technology-oriented development projects implemented in communities, people can collectively become agents of change rather than being mere recipients of transferred technology. Through participation in collective technology-oriented projects, people can re-negotiate the distribution of resources, challenge inequalities and claim their rights, hence gaining individual, but also collective agency.

8.4 Technologies for Freedom

What are the differences between the Case 1 (Chucura), Case 2 (Charía) and Case 3 (Nueva Alianza) projects if all three got a rural area electrified? As we have seen in the three cases, a technological project was implemented in a poor community, using an appropriate technology: a micro hydro power plant. It is a reliable and robust technology, which does not usually cause too many maintenance problems. But, as we argue in this chapter, technology is not the only factor that ensures the success of technological intervention. Chucura inhabitants were not motivated or involved in the electrification project. In Charía's case study we see an expansion of individual capabilities, but hardly can we say that there was an expansion of collective agency of their members. We think that the project in Nueva Alianza is the really successful one, as there was a significant change following the project. This change results, in terms of the capability approach, in an enhancement of individual capabilities, but also in a strong expansion of individual and collective agency, promoting the ability of the inhabitants of Nueva Alianza to shape their future and pursue the goals they value.

We propose to use the capability approach for re-thinking technology in relation to human development. The new conceptualisation we propose allows technology-oriented development projects implemented in small communities to make an optimal contribution to the human development of its inhabitants. As we already mentioned, from the capability approach perspective, development deals with expanding people's freedom and choices (capabilities), making people more capable of shaping their future or influence the world (agency). This only makes sense under principles of equity, productivity, empowerment and sustainability (the basic principles of human development). As we referred to above when explaining the vision of the UNDP, expanding the capabilities means *de facto* expanding people's agency. However, our proposal is that interventions should explicitly prioritise and promote the expansion of agency. That is, technology projects, when properly set up and implemented, serve as vectors to expand the freedoms and choices for people, but also to enhance their ability (individually and as a group) to pursue goals they consider valuable. Under these conditions we can speak of *Technologies for Freedom (T4F)*.

8.4.1 What Is New in T4F?

Applying in depth the Capability Approach involves going beyond the vision of the UNDP HDR 2001 or the Appropriate Technologies approach. And that is the main challenge of developing T4F: re-thinking and developing technological processes that incorporate, from conceptualisation to implementation, an intention to promote human development.

The question is: How can a technological aid project serve as a vector to expand the freedoms of community members? How can the project enhance the ability of the members to pursue the goals they consider valuables?

To implement the capability approach in technology-oriented development projects has some practical implications. Although further research is needed, in this section we present some considerations that should be considered if the objective of the development project is to expand the capabilities of people, while *enhancing* the expansion of the agency (individual and collective).

In this regard, under the T4F perspective, the technician's role is not to implement a development project, but to facilitate a *development process led by the community*. Therefore, the decisions to be taken should not be presented as purely technical choices. The technician should present and facilitate the technical issues and different options as objectively as possible.

This is only possible when *inclusive spaces for deliberation and participation* are generated, and when people are motivated and involved in the process of choosing, designing and regulating technology. We understand design not in its narrow sense of making difficult technical calculations, but in its broader sense of valuing and choosing between different technical aspects. This task is usually reserved for external experts that may (but need not) pay attention to the interests of the community. But it is possible to involve the community in this stage of the project more directly. Workshops can be done where experts share their knowledge with the community, explaining the different design options, their consequences, the criteria to be taken into account, etc. And with good and objective facilitation, the *learning process* will not only travel from experts to the community but also from the community to the experts, who will better understand the community views. Thus, there will be a basis of information available for the *community to be able to engage in this decision making process*, and their views and values will be taken into account (not only the technicians' ones).

Under the T4F vision, each technology must be developed or adopted and adapted in interaction with the community or by the community itself. The technological artefacts (products, equipments, etc.) and the organizational processes and relationships are *ends* of the interventions; but they also represent the *means* that allow people to do and achieve whatever goals or values they regard as important (individual and collective capabilities), *enhancing the ability* of the community to help themselves to make changes happen (individual and collective agency).

As we have seen above, this only makes sense under the *principles of human development* according to the capability approach. Interventions should ensure equity (participation of men and women, young and old, households affected by the HIV/AIDS pandemic and those that are not, etc.), sustainability (not only environmental, rather, a sustainable collective empowerment process that persists in time), productivity (improving people's knowledge and power to make technology choices and expanding potentialities and capabilities of people to ensure income) and empowerment (enhancing the power of people to make changes happen).

	Appropriate Technologies (AT)	Technology for Human Development (UNDP vision)	Technologies for Freedom (T4F)
Focus on	Ends	Ends	Processes
Generation of knowledge in community	Not a target	Not a target	Yes
Technology transfer process	Top-down	Top-down	Bi-directional: Bottom-up and Top-down
Community participation	Yes	Yes	Yes
Role of technician	Essential (to adapt technology)	Intermediary ("seller" or offerer of innovations)	Secondary (to present and facilitate options)
People involved and motivated	Not necessary	Not necessary	Necessary
Expansion of capabilities	Not a priority	Yes	Yes
Expansion of agency (individual and collective)	Not a priority	Not a priority	Yes

Fig. 8.5 Comparison of different approaches in community technology-oriented development projects

8.4.2 TF4 Versus Other Approaches

Some of the features of the AT are shared by T4F, as community participation, local and intensive labour force, use of natural resources and respect for local culture and environment. It also agrees with the UNDP vision that technological innovation and economic growth are important to build human capabilities.

Nevertheless, T4F differs from both positions in two issues of great importance. Firstly, *it focuses mainly on the process* instead of stressing the results and products of the interventions. And secondly, community *people* play the *central role* in the generation and dissemination of knowledge. The type of participation in T4F projects differs from other approaches. In the AT and UNDP vision of technology transfer, the community is just informed or consulted by technicians, while under the T4F the community takes the central stage: it is fully involved from the start. Technicians actively encourage the local knowledge, informal research and development systems and facilitate community's experimentation. In the best of cases, the community even relies on their own experimentation and there is no organized communication with technicians and extensionists.

In the next figure we compare, simplistically, T4F with other approaches in community technology-oriented development projects.

Figure 8.5 analyses the core of the different visions on technology. However, in real life there will be projects that cannot be solely classified as AT, UNDP or T4F, as they will contain features of two or three categories.

8.5 Conclusion

In this chapter we have explored the contributions of the Capability Approach to technological aid projects implemented in small communities or villages. We have presented three development projects implemented in rural villages, funded by similar donors, and obtain the same results (*ends*) in the space of resources – a sufficient amount of energy for the communities. Nevertheless, in the first case – Chucura-, the project of the power plant was seen as merely a technological project, and as a consequence no attention was paid to social and relational issues or local processes (*means*). The project thus had little impact on the development process in the community and did not contribute at all to the empowerment of its members. In the second case, Charía, we can see an expansion of individual capabilities, but it is difficult to say that there was a strong expansion of agency of their members. But in the third case, Nueva Alianza, the community participated in all the stages of the process, confirming their engagement with the electrification project. Through community participation, the power plant is working properly and has served as a driving force for the community to become involved in other new projects, to claim their rights, to engage in rural movements and to support other communities.

We contend that technology-oriented development projects can be vehicles to expand the freedom of people. The technological artefacts (products, equipments, etc.) and the organizational processes and relationships are *ends* of the community interventions; but they also represent the *means* that allow people to do and achieve whatever goals or values they regard as important (capabilities), enhancing the ability of the community to help themselves to make changes happen (agency). And, what is more important, people can *collectively* become agents of change rather than being mere recipients of aid. Through participation in collective technology oriented projects, people can re-negotiate the distribution of resource, challenge inequalities and claim their rights, gaining individual, but also collective agency.

We have also introduced a new conceptualisation of technology that incorporates, from conceptualisation to implementation, an intention to promote human development. Thus, we have presented *Technologies for Freedom* (T4F) as the community-driven technological processes intended to expand people's freedoms and choices, but also to enhance their ability (individually and as a group) to pursue goals they consider valuable. Some features of T4F community development projects have been pointed out, stressing the importance of participation and motivation, knowledge creation and capacity building, as well as the collective agency processes of the communities.

Some characteristics of T4F can be extrapolated to any intervention. However, in this work we have focused on technology projects because it is something incipient and emerging in this field. Technology projects have traditionally paid little attention to community participation and empowerment of people based on the premise that technology requires high level of knowledge, available only to external technical experts. But we believe that another way to understand technological interventions is possible.

This work focuses only on technological projects carried out in communities or villages. It does neither pay attention to technology transfer processes, nor to applications that people make of technology. It does not focus on global issues that affect poverty. So, aware as we are of the limitations of this study, more research must be carried out in order to establish the relationships between local and global or the uses people can make of technology. However, we strongly believe that agency expansion (individual and collective) can be an effective means for people's substantive freedoms. This agency expansion must be prioritized in technology aid projects and play an important role in the development agenda.

Acknowledgments We are very grateful to Alex Frediani, from the Development Planning Unit of the UCL, for his valuable comments to the different drafts of this chapter. We wish to thank as well Ilse Oosterlaken and her colleagues at Delft University for their efforts to introduce discussions on technology into the Human Development and Capability Approach community. Finally, we really appreciate the support of our University, particularly the Centro de Cooperación al Desarrollo, to carry out our research.

References

Alkire, S. (2002). *Valuing freedoms: Sen's capability approach and poverty reduction* (Queen Elizabeth house series in development studies). Oxford: Oxford University Press.
Alkire, S. (2005). Subjective quantitative studies of human agency. *Social Indicators Research, 74*, 217–260.
Alkire, S. (2008). Concepts and measures of agency. In K. Basu & K. Ravi (Eds.), *Arguments for a better world: Essays in honor of Amartya Sen. Vol. I: Ethics, welfare and measurement* (pp. 455–474). Oxford: Oxford University Press.
Alkire, S., & Deneulin, S. (2009). Human development and capability approach. In S. Deneulin, L. Shahani, S. Deneulin, & L. Shahani (Eds.), *An introduction to human development and capability approach* (Vol. 22). London: Earthscan.
Ballet, J., Dubois, J. L., & Mahieu, F. (2007). Responsibility for each other's freedom: Agency as the source of collective capability. *Journal of Human Development and Capabilities, 8*(2), 185.
Behari, B. (1976). *Rural industrialization in India*. New Delhi: Vikas Publishing House.
Brandão, F. C. (2001). *Programa de apoio às tecnologias apropiadas - PTA: Avaliação de um programa de desenvolvimento tecnológico induzido pelo CNPq*. Brasilia: UnB.
Chambers, R. (1997). *Whose reality counts? Putting the first last*. London: Intermediate Technology.
Comim, F., & Kuklys, W. (2002). *Is poverty about poor individuals?* Paper presented at 27th general conference of the International Association for Research in Income and Wealth, Djurham.
Crocker, D. (2008). *Ethics of global development: Agency, capability, and deliberative democracy*. Cambridge: Cambridge University Press.
Dagnino, R. (1976). *Tecnologia apropiada: Uma alternativa?* Brasilia: UnB.
Dagnino, R., et al. (2006). Política científica e tecnológica e tecnología social: Buscando convergência. In *Registro do forum nacional da RTS*. Brasilia: Abipti.
Deneulin, S., & Stewart, F. (2001). *A capability approach for individuals living together*. Paper presented at Justice and Poverty: Examining Sen's Capability Approach, Cambridge.
Deneulin, S., & Stewart, F. (2002). Amartya Sen's contribution to development thinking. *Studies in Comparative International Development, 37*(2), 63.

Dubois, A. (2006). El enfoque de las capacidades. In Alejandra Boni, Agustí Pérez-Foguet, & Intermon-ISF (Eds.), *Construir la ciudadanía global desde la universidad*. Barcelona: Publicaciones ISF.

Dufumier, M. (1996). *Les projets de développement agricole, manuel d'expertise*. Khartala: Broché.

Evans, P. (2002). Collective capabilities, culture and Amartya Sen's development as freedom. *Studies in Comparative International Development, 37*(2), 54.

Fernández-Baldor, Á., Hueso, A., & Boni, A. (2009, September 12–14). *Technologies for freedom: Collective agency-oriented technology for development processes*. Paper presented at the Human Development and Capability Approach Conference. Lima.

Fukuda-Parr, S. (2003). The human development paradigm: Operationalizing Sen's ideas on capabilities. *Feminist Economist, 9*(2–3), 301–317.

Gasper, D. (2007). What is the capability approach?: Its core, rationale, partners and dangers. *Journal of Socio-Economics, 36*(3), 335–359.

Goodman, L. J. (1976). *Appropriate technology study: Some background concepts, issues, examples and recommendations* (Vol. IV). Honolulu: University of Hawaii.

Herrera, A. (1983). *Transferencia de tecnología y tecnologías apropiadas: Contribución a una visión prospectiva a largo plazo* (Tesis doctoral, Unicamp. Campinas (Brazil)).

Hueso, A. (2007). *Estudio sobre el impacto social, económico y ambiental de pequeñas centrales hidroeléctricas implantadas en comunidades rurales de la paz, bolivia*. Valencia: Universitat Politècnica de València.

Ibrahim, S. (2006). From individual to collective capabilities: The capability approach as a conceptual framework for self-help. *Journal of Human Development and Capabilities, 7*(3), 397–416.

James, B. (1995). *The impacts of rural electrification: Exploring the silences*. Cape Town: Energy Development and Research Center.

Kumar, K. (1993). Mohandas Karamchand Gandhi. *Perspectivas: Revista Trimestral De Ed-ucación Comparada, XXIII*, 535–547.

Leach, M., & Scoones, I. (2006). *The slow race. Making technology work for the poor*. London: Demos.

Nussbaum, M. C. (1987). Nature, functioning and capability: Aristotle on political distribution (Working Papers 1987/31). Helsinki: UNU-WIDER.

Nussbaum, M. C. (2000). *Women and human development: The capability approach*. Cambridge: Cambridge University Press.

Pérez-Foguet, A., Lobo, M., & Saz, Á. (2005). *Introducción a la cooperación al desarrollo en las ingenierías: Una propuesta para el estudio*.Associació Catalana d'Enginyeria Sense Fronteres.

Robeyns, I. (2003). Sen's capability approach and gender inequality: Selecting relevant capabilities. *Feminist Economics, 9*(2, 3), 61.

Robeyns, I. (2005). The capability approach: A theoretical survey. *Journal of Human Development, 6*, 93–117.

Schumacher, E. F. (1973). *Small is beautiful. Economics as if people mattered*. New York: Harper and Row.

Sen, A. (1979). *Sobre la desigualdad económica*. Madrid: Editorial Crítica.

Sen, A. (1982). *Poverty and famines: An essay on entitlements and deprivation*. Oxford: Clarendon Press.

Sen, A. (1989). Development as capability expansion. *Journal of Development Planning, 19*, 44–58.

Sen, A. (1985). Well-being, agency and freedom: The Dewey lectures 1984. *The Journal of Philosophy, 82*, 169–221.

Sen, A. (1992). *Inequality reexamined*. New York/Oxford: Russell Sage Foundation/Clarendon Press.

Sen, A. (1999). *Development as freedom*. New York: Oxford University Press.

Sen, A. (2002). *Rationality and freedom*. Cambridge: Belknap.

Shiva, V. (2009). *The seed and the spinnig wheel: The UNDP as biotech salesman.* [cited June, 09 2009]. Available from http://www.poptel.org.uk/panap/latest/seedwheel.htm

Stewart, F. (2005). Groups and capabilities. *Journal of Human Development, 6*(2), 185–204.

Ul Haq, M. (1995). *Reflections on human development.* Oxford/New York: Oxford University Press.

United Nations Development Programme. (2001). *Human development report 2001: Making new technologies work for human development.* New York: Oxford University Press.

United Nations Development Programme. (2005). *Human development report 2005: International cooperation at a crossroads: Aid, trade and security in an unequal world.* United Nations Development Programme.

Walker, M. (2006). *Higher education pedagogies. A capabilities approach.* Berkshire: Open University Press.

Watts, M., & Bridges, D. (2006). Enhancing students' capabilities? UK higher education and the widening participation agenda. In S. Deneulin, M. Nebel, & N. Sagovski (Eds.), *Transforming unjust structures.* Dordrecht: Springer.

White, L. J. (1974). *Appropriate technology and a competitive environment: Some evidence from Pakistan* (Discussion Papers (46), ITS).

Chapter 9
Technology Choice in Aid-Assisted Parliamentary Strengthening Projects in Developing Countries: A Capability Approach

Malik Aleem Ahmed

9.1 Introduction

Citizens in many developing countries, apart from being economically poor, are also poor with respect to information access and availability (Ahmed 2010). Many people in developing countries do not have access to information and communication technologies (ICTs) and consequently timely information, even if the infrastructure is available (Dada 2006). In other words, citizens do not have the proper access to information resources, including parliamentary information resources. As a result, citizens in many developing countries are not benefiting from the so-called information revolution. Citizens do not have the proper tools and freedoms to access, use, and disseminate information. Aid assisted parliamentary strengthening initiatives should aim at designing an environment for expanding the information capabilities of the citizens, members of parliaments, parliaments as institutions and other stakeholders (Ahmed 2011b).

The aim of this chapter is to explore the added value of the capability approach for selecting suitable information and communication technologies and systems during aid assisted parliamentary strengthening projects (AAPS-projects). Considering this aim and the complex nature of systems implementation, a desk research methodology has been adopted. I also analyze some empirical facts in order to support the arguments for implementing parliamentary telecasting systems in developing countries during AAPS-projects. I discuss that if decision makers need to choose a system from the available options then the capability approach perspective can assist them in the process.

M.A. Ahmed (✉)
Department of Values and Technology, Delft University of Technology,
Delft, The Netherlands
e-mail: m.a.ahmed@tudelft.nl

The chapter is structured as follows. In Sect. 9.2, I start with a brief introduction on the capability approach, the role of ICTs and introduce the reader to the notion of information capabilities. I describe the connotation of AAPS-projects in Sect. 9.3, after discussing the inability of the parliaments in developing countries to utilize ICTs. In Sect. 9.4, I indicate that often the emphasis of e-parliament projects is on Internet-related systems for expanding the information capabilities of parliamentary stakeholders in developing countries. I confer that there is a need to consider different possible means and alternatives. I use the concepts of the capability approach and the notion of information capabilities to argue that selection of technology should be context-dependent, based on the conversion factors. Social and economic factors, budget, and technical resources also play a role in the selection process. I present a scenario where telecasting systems can be preferred over Internet related systems from the perspective of the capability approach. However, parliamentary telecasting systems also have some limitations from an information capabilities perspective, as identified in Sect. 9.4.4. In Sect. 9.5 I present some arguments that are sometimes made for *not* investing in parliamentary telecasting initiatives. These arguments are based on the economic development or utility maximization perspective. I discuss that some of these arguments seem to lose force if the initiatives are viewed through the lens of the capability approach.

In this chapter, I present three short illustrative cases from different AAPS-projects, which show that parliamentary telecasting systems can be implemented by taking different paths depending upon the conversion factors especially upon some aspects of environmental conditions. Case 1 illustrates that in certain scenario the public-private or public-public partnership can be sought for sustainable operations of parliamentary telecasting systems. Case 2 illustrates how a parliamentary telecasting initiative can be started. Case 3 illustrates the efforts for overcoming the opposition to parliamentary telecasting initiatives. The chapter finishes with the concluding remarks.

9.2 The Capability Approach, ICT and Information Capabilities

The underlying theory behind the concept of information capabilities is the capability approach (see Sen 1980, 1985, 1999, 2002; Nussbaum 2000; Alkire 2002; Robeyns 2003, 2006). The capability approach stresses that the focus of development should be on human development, agency, well-being, and on providing freedoms to the people instead of only on economic development or utility maximization (Sen 1999). Economic development is considered as one of the means for human development (Sen 2002) along with political freedoms, social opportunities, transparency guarantees, and protective securities (Sen 1999). The capability approach focuses on what people are realistically able to do and to be (Nussbaum 2000) in a given context.

Recently attention has been given to technology in general and Information and Communication Technology (ICT) in particular as potential means for the expansion of human capabilities (e.g. Ahmed 2010, 2011b; Garnham 1999; Johnstone 2007; Kleine 2010b; Madon 2005; Mansell 2001; Musa et al. 2006; Oosterlaken 2009; Zheng 2007).

There have been several attempts to operationalize the capability approach in the field of ICT for development (e.g. Alampay 2006; Kleine 2010a, b; Zheng 2007; Zheng and Walsham 2008). Some studies – for example, (Madon 2005; Prakash and De' 2007) – have used the concepts of the capability approach for analyzing the outcomes of e-government projects. However, there is very little literature available on using the concepts of the capability approach in e-parliament initiatives and especially in AAPS-projects. This paper is also an attempt to address that topic.

Functionings and *capabilities* are the core concepts in the capability approach. *Functionings* are things a person may value doing and being (Sen 1999). Functionings can vary from very general (like living a healthy life) to very specific (for example, acquiring specific information). Capability has been defined in more than one ways, but for the purposes of this chapter, *capability* means the positive freedom to do things and/or to be which the person values (e.g. see Sen 1999; Nussbaum 2000). In other words, the term 'capability' refers to the positive *freedom of realizing various functionings*. This definition of *capability as* the freedom to do things and/or to be *which the person values* also points towards the importance of choice. A person should have access to alternative means, resources, and infrastructure in order to be able to realize different functionings. Similarly, from the agency perspective, an agent should have different choices available to act as he/she would like to.

Information is a very important resource and has an instrumental value for achieving or doing important things in life. We can say that people value acquiring, using and disseminating information because it can lead them in making decisions for doing things and consequently for living the life they value. ICTs can facilitate users in the instrumentally important functionings of accessing, using, and disseminating information. *Information capabilities* refer to the freedoms of realizing these functionings of (1) acquiring, (2) using and (3) disseminating information. In case of ICT and especially the Internet, information capabilities can also be referred as e-capabilities. E-capabilities can be thought of as information capabilities created through e-means, tools, and systems. Information capabilities can contribute to many ultimately valuable capabilities of citizens, capabilities that enable them to live the lives they have reasons to value. For example, information capabilities are often needed for the functionings of communicating and interacting with others. They also facilitate making different decisions and acting on them. More indirectly, information capabilities are instrumentally important for other kind of capabilities including being able to lead a healthy life, receiving education and earning an income (e.g. see Mansell 2001). Nussbaum (2000) has identified a list of ten essential capabilities that democratic states should support.[1] Information capabilities can contribute to all the essential capabilities on her list. Of particular relevance, in our case, are the capabilities related to her category of control over one's political environment. With reference to AAPS-projects they also assist in stimulating political participation and engagement. Expanded information capabilities might put the citizens in the better position to

---

[1] The list includes; (1) Life, (2) Bodily Health, (3) Bodily Integrity, (4) Senses, Imagination, and Thought, (5) Emotions, (6) Practical Reason, (7) Affiliation, (8) Other Species, (9) Play and (10) Control over one's Environment (a) Political and (b) Material.

monitor political abuses. Similarly, we can say that information capabilities enable the citizens to exercise their rights in the right direction – for example, political right of voting. In this case, when the citizens have certain information then they would be in a better position to exercise their voting rights depending on the quality and truthfulness of information. An expansion of information capabilities can leads towards open, transparent and accountable parliaments.[2]

The notion of 'information capabilities' differentiates from merely capacities or skills. A person may have the skills to perform a certain functioning, but the question that a capability approach theorist is interested in is "does that person has the positive freedom or real opportunity to realize that functioning?" The term 'realizing' has been used in the sense of 'realizing a plan'. There might be a plan, but to realize that plan into reality some conditions have to be met. Similarly, to realize a potential functioning into an actual functioning, some internal and external conditions have to be met. Just having skills – like computer skills or reading skills – may not be enough. One of the central insights of the capability approach is thus that means might not be useful if the contextual environment is not (sufficiently) suitable. As Zheng (2007) notes, "… individual variations, as well as structural differences in society, are important factors to be taken into account in evaluating development initiatives." *Freedom of realizing* implies that certain background conditions have to be provided in order for realizing the *potential functioning* into an actual one. We can note, following Sen's footsteps, that the freedom to realize a potential functioning depends on background conditions, also called conversion factors, such as *personal characteristics*[3] – for example, literacy –, *social conditions* – for example, social incentives, punishment, fears and desires – and *environmental conditions* – for example, climate, infrastructure, resources, and public goods. The emphasis of this chapter is on environmental conditions, especially concerning tools and infrastructure provision. Designing an environment for *information capabilities expansion* through ICTs, means using ICTs for creating more opportunities and freedoms for the stakeholders for acquiring, using, and disseminating information.

From an information capabilities perspective, citizens should be empowered in such a way that they have alternative resources and public goods available to choose for accessing the desired information, communicating, sharing and disseminating the information. A citizen can choose not to function i.e., he/she might choose of not performing the function of acquiring and disseminating information, but if he/she chooses to do so then states have the responsibility to make sure that there are sufficient means available for the citizen. Citizens should have the freedom to use those means and choose from alternative means.

---

[2] Here the assumption is that the right information and not controlled or misleading information is shared with the citizens. The availability of correct information is necessary. If misleading information is available, then citizens would make decisions based on that information.

[3] One can argue that personal characteristics are important for functioning even if the freedom to realize that functioning is not available in a given context. Personal characteristics can be thought of as the indirect background condition for realizing a functioning.

Though the major emphasis of the capability approach is on a person, many of the concepts of the approach can also be applied to the institutions.[4] For realizing a citizen's capability of acquiring information – e.g. with respect to the parliaments – the information has to be available in the first place. This means that parliaments should have the capabilities to gather, process and share the information with the citizens. Similarly, the purpose of citizens' functioning of dissemination of information to the parliaments is fulfilled when the parliaments have the freedoms for realizing their functioning of acquiring and using (processing) that information. This also implies that the information capabilities of the citizens with respect to the parliaments are dependent on the information capabilities of the parliaments. Therefore, to expand the information capabilities of the citizens with respect to the parliaments, the information capabilities of the parliaments need to be expanded. Aid-Assisted parliamentary strengthening projects are one of the ways to design an environment of information capabilities expansion for the parliaments, which in turn could result in expanding the information capabilities of the citizens and other parliamentary stakeholders. In later sections, I discuss the importance of information capabilities of the parliaments more extensively.

9.3 Aid-Assisted Parliamentary Strengthening (AAPS) Projects

In one of her reports USAID wrote the following:

> Democracy can only be realized when legislators have the will, ability, and information to make decisions that reflect the interests and needs of society. Likewise, the people must have the will, ability, and information to transmit their needs and interests to the legislature, to evaluate the performance of legislators and their parties, and to reward or sanction their action. (USAID 2000, p. 7)

Political freedoms in the form of democratic arrangements help to safeguard economic freedom and the freedom to survive (Sen 1999). Parliaments are the central institutions representing the people in the democratic arrangements. Technology enables the parliaments to realize the values of transparency, accessibility, and accountability (Griffith and Casini 2010). Many parliaments have started realizing that to gain respect in the eyes of citizens of their countries, they need to exchange information with the citizens and make their processes and functions effective and transparent:

> […] parliaments across the world are actively seeking to respond to the challenges of the present age. Mostly they are doing so by improving their ongoing procedures… (Beetham 2006, p. 9)

However, parliaments in many developing countries perform poorly and tend to be closed institutions (Ahmed 2011a). One of the reasons for poor parliamentary

---

[4] Here I would like to share that the expansion of the information capabilities of the institutions/organizations is important in so far as they contribute to the citizen's information capabilities.

performance is that parliaments in many developing countries have not been able to employ ICTs to improve their main functions and processes. In many instances, they are not able to exchange timely information with the citizens and other stakeholders. Many of the e-government initiatives [including e-parliaments initiatives] in developing and transitional countries are either total failures or partial failures (Heeks 2002, 2003). Yet:

> [...] transparency, accessibility and accountability, as well as people's participation in the democratic process, largely depend on the quality of information available to members of parliaments, parliamentary administrations, media and the society at large, and on citizen's access to parliamentary proceedings and documents. Both can be improved through ICT applications [for example, parliamentary telecasting systems], which in turn could dramatically strengthen the policymaking process. (Griffith and Casini 2010)

There is an urgent need for developing, strengthening and opening up the parliaments in developing countries, as Griffith and Casini (2010) argue convincingly. In other words, there is an urgent need to design an environment for the parliaments for expanding their information capabilities, which could result in the expanded information capabilities of the Members of parliaments, citizens and other stakeholders.

As mentioned earlier, information capabilities of the parliamentary stakeholders, including the parliaments, are interdependent to some extent. For example, the information capabilities of citizens are dependent on the information capabilities of the parliaments. The capability of the parliament to access raw data, convert it into useful information, and share it with the citizens translates that the citizens can access the information and then use that information to make decisions. Similarly, the information capability of the citizen to send [disseminate] the information to the parliaments would be beneficial if parliaments have the information capability to use [process] the information for making future decisions.

It may be argued that many poorer countries do not have the required money, technical knowledge, expertise, and the will to strengthen their parliaments. This is where developed western countries come into the picture. Many western countries are motivated to assist new and transitioning democratic countries. One way to do so is to lend or give money to the developing countries. However, many of us would agree that this may not be the best solution, as there is lack of expertise and honesty in the government sectors of developing countries. One of the alternative ways is for donor countries to channel funding and technical assistance for the implementation of project through development organizations.[5]

Projects aiming at strengthening the parliaments, in which the international stakeholders like aid donors and international development organizations are involved and they provide financial and technical assistance to the parliaments in developing countries, are called aid assisted parliamentary strengthening projects. The United States Agency for International Development (USAID), the Canadian International Development Agency (CIDA) and the UK Department for International Development (DFID) have been providing support to parliaments of many countries

---

[5]The assumption here is on the good intentions of the donors, international development organizations, and local implementing parties.

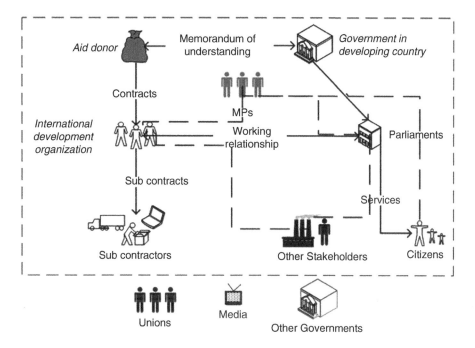

Fig. 9.1 Implementing parties and stakeholders of AAPS-projects

in the bilateral mode of assistance. Similarly, the United Nations Development Programme (UNDP) has been working for parliamentary strengthening as institutions of governance in many countries (UNDP 2010). The World Bank is also active in this area. Many stakeholders are directly involved or indirectly affected by these kinds of projects as shown in Fig. 9.1.

If AAPS-projects involve some major ICT interventions, then we might call them aid assisted *e*-parliament projects or interventions. We could assert, based on Capability approach, that the implementing parties should focus on providing entitlements and empowerment to the parliamentary stakeholders. Empowering the stakeholders could result in more freedoms for the parliamentary stakeholders to realize the potential functionings into actual ones i.e., empowerment could result in capabilities expansion of parliamentary stakeholders. For example, if citizens are empowered to be able to access information, then it means their information capability of access of information is expanded.

9.4 Choosing a Technological System: The Capability Approach

In an ideal situation, a combination of technologies and systems could be used for designing an optimal environment for information capabilities expansion of the parliamentary stakeholders during AAPS-projects. I have discussed elsewhere

(Ahmed 2011b), that different ICT initiatives – for example, parliamentary website development initiatives, webcasting systems, broadcasting systems, printing capacity improvements, reporting centers enhancement, media centers upgrading, and resource centers expansion – can be undertaken to assist in the process of expanding the information capabilities of the citizens, members of parliaments and other stakeholder groups via AAPS-projects.

Sometimes implementing parties have to choose one technological solution from the many tentative options. The reasons for this limitation of choosing one technological solution can be many, including time limitations, financial restrictions, resources constraints and scope limitations of an aid assisted parliamentary strengthening initiatives. Too much emphasis is then often put on using the Internet and Web-related tools and technologies for opening up and strengthening the parliaments (e.g. Ahmed 2008; Griffith et al. 2008; Griffith and Casini 2010; IPU 2009; Leston-Bandeira 2007). Internet related systems – for example, websites and webcasting systems – can in principle contribute to an environment for expanding the information capabilities of parliamentary stakeholders. However, as indicated in Sect. 9.2, personal, social and environmental conversion factors have to be taken into account in order to turn the *potential functionings* into actual ones. Employing conventional media technologies, which are widely and easily accessible, seems a good alternative. Those without the means to access Internet services then also have ways of acquiring parliamentary information. One example falling within the realm of conventional electronic media are parliamentary telecasting systems. Parliaments are the institutions that have a legitimate interest in allowing broadcasters to inform the public about their work (Johnsson 2007). In the following section, I investigate in more detail why parliamentary telecasting initiatives might be chosen, as opposed to Internet related systems in a developing country. The discussion takes into account the relevant background conditions, conversion factors, i.e. personal characteristics, social conditions, and environmental conversion factors.

9.4.1 Personal Characteristics

It could be argued that the parliamentary websites and web-casting systems could only assist in expanding the information capabilities of those who possess the personal characteristics, e.g., reading, writing, know how about using the Internet and others. To access the information via the TV channels, people do not need such personal characteristics. The literacy level is generally lower in many developing countries, which also translates into the less people being able to use the Internet resources. In those countries where the literacy level is low and many people do not have access to broadband Internet systems – for example, a webcasting system – could also exacerbate the information capabilities inequalities within the country.

Table 9.1 Conversion factors comparison

Conversion factors	Internet related systems	Telecasting
1. Personal characteristics	Skills like reading, writing, typing, using the Internet	Minimum
2. Social characteristics	Fear of using Internet, Social acceptance	Relatively more socially acceptable
3. Environmental characteristics	Access to computers and Internet (in many cases broadband Internet)	Access to TV set

9.4.2 Environmental Conditions

Citizens need resources such as Internet connections, telecom infrastructure, access devices like computers, to acquire, use, and disseminate information via the Internet technologies, systems, and applications. The stakeholders of AAPS-projects should keep in mind that access to Internet, and especially broadband access, is limited in many developing countries. The environmental conditions perspective forces the stakeholders to pay attention to the Internet access status in the specific developing country in question. In many of the developing countries the Internet penetration and usage is relatively low than world average of 27.8% (e.g., see IWS 2010). For example, in the case of Pakistan, less than 1% of population has access to broadband Internet (PTA website). Whereas 50% of the population has access to a TV set. In this specific scenario, where more citizens have access to TV than to computers and Internet, a parliamentary TV channel looks a more attractive option to disseminate parliamentary information. In this way, the information capabilities of as many citizens as possible will be expanded.

9.4.3 Social Conditions

Internet is relatively a new technology. Fear of using the Internet still exists in many poorer countries, especially among the older population. In certain cases people of the poorer countries are afraid of the Internet and think of it as dominated by Western cultural values. People in many of the poorer countries are socially comfortable with the telecasting and TV system. It is assumed, based on the yearly stats on Internet World Stats website (IWS 2010), that, at least in the near future, many citizens will find it easier and more socially acceptable to access the parliamentary information from the TV channel than the Internet and the Web.

Table 9.1 presents a summary of the different conversion factors discussed. It shows that parliamentary telecasting system seems a more suitable option as compared to Internet related systems.

9.4.4 Limitations of Parliamentary Telecasting Systems from Information Capabilities Perspective

The conversion factors perspective shows that parliamentary telecasting system is the better option to choose if the stakeholders have to invest in just one of the available options. However, parliamentary telecasting systems, in contrast to Internet related systems, have some limitations from an information capabilities perspective.[6] For example, telecasting can be a very good means for expanding the information dissemination capability of a parliament, but not for expanding its information acquiring capability. We should also keep in mind that telecasting is a one-way dissemination of information i.e., citizens have very limited information capabilities to disseminate information and hence communicate with the parliaments in this way. Moreover, the information on the TV channels is available only at a fixed time. If a citizen misses a program, there are very limited ways to acquire that information through this option. New technological developments could make it possible for parliaments to get the feedback from the citizens through parliamentary TV channels, but probably not in the near future.

Whereas, in case of the web, the communication and information flow is two ways. Citizens have tools to disseminate information and the information on the web stays available for a much longer period. Citizens can access it at any time. Thus, there can be different ways to achieve the same objective. A more in-depth analysis is needed on a case-by-case basis to determine which alternative would be best. Cost and technical resources required for establishing and sustaining a website compared to the cost and resources required for of establishing and sustaining the parliamentary telecasting system are lower.[7] Another aspect to see from the information capabilities perspective is that the information on the Internet is easily searchable. Moreover, other factors including political willingness, social constraints, ease of use, number of people, durability of information plays a role in reaching the decision.

The information capabilities perspective, in combination with attention for conversion factors, can change the final choice of system and technology.

9.5 A Response to Arguments Against Parliamentary Telecasting Initiatives

Many arguments have been made at different platforms for not investing in ICT initiatives in the public sector institutions of developing countries. Some arguments relate to the challenges and issues in the developing countries. Some of the general

---

[6]This can be considered as an advantage of taking Information capabilities perspective on e-initiatives.

[7]However, in case of other internet-based systems – for example, webcasting systems – the same analogy can be applied for in favor of telecasting systems.

challenges and problems include poor infrastructure (telecom, broadband Internet), lack of skilled ICT-personnel, poor leadership style, bureaucracy, poor attitudes, lack of coordination and continuously changing ICT policies (Gichoya 2005). Other problems include the technocratic approach by donors, failure to understand social and political phenomena, failure to develop local ownership, failure to use local leadership, a harsh local environment, little or no accountability and a low level of administrative capacity (Schacter 2000).

Some of the challenges involved in using technologies in the parliaments include lack of familiarity of members with technology, lack of knowledge of citizens about the technology, limited access to the technology and insufficient understanding of the citizens with the legislative processes (Griffith and Casini 2010). "[…] there are many parliaments that appear to be adopting "bits and pieces" of technology", say Griffith and Casini (2010, p. xi), "but without a coherent and sustained vision. These legislatures are failing significantly to reach the full potential ICT can offer to strengthen the institution; they may, in fact be falling behind."

At least four arguments can be found in the literature for not investing in the system for telecasting of the sessions of parliaments in the developing countries. They are categorized as: (1) Economic and financial feasibility argument, (2) Unpopular in developed societies argument, (3) Low or no demand argument and (4) Parliaments not ready argument. These arguments are generally supported from the economic or utilitarian perspective. In the following sub-sections, I will respond that these arguments lose force when seen through the theoretical lens of the capability approach.

9.5.1 Economic and Financial Feasibility Argument

TV transmissions are expensive, and most of parliamentary channels in developed countries are financed by public funds (IPU 2007). Financial and technical resources are scarce in the parliaments of developing countries (e.g. see Hudson and Wren 2007). Public funds are not readily available and often times are inadequate. Many times an argument is presented that the parliamentary TV channels and telecasting system for parliamentary sessions are not economically feasible, as financial resources are required to establish and sustain the system (IPU 2007). A consequent argument is that parliamentary TV channels cannot compete with the private TV channels and hence they cannot generate enough revenues (IPU 2007) to sustain the operations. These arguments are made from the economic perspective.

Response: In the capability approach literature, it has been discussed that "The state and the society have extensive roles in strengthening and safeguarding human capabilities" (Sen 1999, p. 53). Sometimes e-parliament initiatives such as establishing a parliamentary telecasting system may not seem useful for monetary or political reasons [or for other reasons] but as (Nussbaum 2000) stated that, in order to be doing what they should for their citizens, states must be concerned with all the capabilities, even when these seems not useful for economic growth. Even if stakeholders feel that designing and sustaining a parliamentary telecasting system would be costly, they should find alternative ways to support and sustain the initiative as shown by the following case.

Case 1 : If the stakeholders feel that, the parliaments would find it difficult to sustain the system due to financial and technical resource scarcity then *public-private partnership* or parliament-public-sector partnership can be sought. In Ghana a public-private partnership was set up to cover plenary sessions and broadcast plenary meetings and committee hearings live on radio (IPU 2007). This solution is suitable for those developing countries where parliaments have limited financial resources and many people do not have access to televisions.

An example of parliament-public-sector partnership was envisioned in Afghanistan by the Support to the Establishment of the Afghan Legislature (SEAL) project implemented by the UNDP. SEAL facilitated a draft agreement between the Parliament and Radio & Television Afghanistan for recording and telecasting two hours of parliamentary news and proceedings every day in 2007. It was believed that the efforts would reinforce the visibility of the Assembly as a representative accountable, and transparent institution; thereby contributing to specific media coverage as well as public debate on parliamentary affairs (SEAL 2007).

9.5.2 Unpopular in Developed Societies Argument

Another argument is that the parliamentary TV channels are not very popular with the citizens in those countries, which have implemented such systems (e.g. Fichtelius 2007); therefore, scarce resources should not be concentrated on developing parliamentary TV channels or invested in designing and establishing parliamentary telecasting systems in developing countries.

If we take the economic perspective then it seems a valid argument. However, taking a capability approach perspective on this argument may make one think in another direction. "The actual achievement of functionings", so Zheng and Walsham (2008) say, "is a result of personal choice to select form the capabilities available, subject to personal preferences, social pressure and other decision-making mechanism." An often-mentioned example in the literature on the capability approach (e.g. Alkire 2002) is that of a person who is voluntarily in the state of malnutrition, because of fasting. This may seem similar to a starving person, who does not have access to food; however the first person has the ability to eat and chooses not to, whereas the second person has the deprivation and would [or could] eat if given a chance. Using the same line of reasoning, in case of establishing a parliamentary telecasting system in a developing country, we could argue as follows. A citizen in a developed country who has chosen not to acquire or communicate information related to parliament may seem similar to a citizen in a poorer country who does not have the tools to acquire or disseminate the information (information deprivation) related to parliaments. However, in the first case, the first person could acquire and exchange information and chooses not to; whereas the second person suffers from a deprivation and would [or could] access, use, and disseminate the information if given an opportunity. Sen (1999) has stressed that the process of development, when judged by the enhancement of human freedom, thus has to include a removal of citizens' deprivation. Yet people should not be forced into certain functionings.

Parliamentary activities are boring, unattractive, and could be considered a "ghetto" segment (Fichtelius 2007). Therefore, many citizens might not be interested in accessing these activities. One way to make parliamentary broadcasting interesting is to broadcast the interviews, debates, and viewpoints of members. As mentioned earlier, citizens and other stakeholders can access the information easily through the parliamentary TV channel from information capabilities perspective. A disadvantage is that the one-way nature of telecasting makes it difficult for the citizens to convey their voices through this media. However, with the advancements in technology and acceptance of Internet Protocol Television (IPTV), it would be possible for the citizens in developing countries to deliver information and give feedback through telecasting option. Nevertheless, in developing countries it could take years for the diffusion and usage of IPTV.

9.5.3 Low or No Demand Argument

Another argument that is made is the low or no demand by the citizens of a specific developing country for the parliamentary telecasting initiative. The capability approach, however, draws attention to the fact that citizens might not demand something because of the 'adaptive preferences' concept discussed by Sen (1999). According to the "adaptive preferences" concept, "deprived people tend to come to terms with their deprivation [...] and they may, as a result, lack the courage to demand any radical change" (Sen 1999). Similarly, we can argue that the information-deprived citizens of the developing countries tend to accept this information deficiency and lack the courage to demand these kinds of initiatives. Decision makers do not realize this and hence do not take the initiative for establishing parliamentary telecasting system. It can result in harming the citizens because, as Group (2008) discussed, that what MPs deliver is partly a function of what citizens understand and expect of them. This includes the relative emphasis that constituents and the broader population place on the different roles that parliamentarians are supposed to fulfill. Demand from the citizens for an active parliament can, in turn, generate demand from an active parliament for effective government (Group 2008). A good example for establishing a parliamentary telecasting system is the initiative taken by Kenya Parliamentary Strengthening Program (PSP).

Case 2: The Kenya Parliamentary Strengthening Program was a jointly funded initiative of USAID and DFID. The Center for International Development, State University of New York, implemented the project from 2000 to 2010. PSP started and completed different initiatives. One of such initiatives was House Live Broadcasting (HLB), which was aimed at the telecasting of plenary sessions and committees activities. The initiative was started to make the Kenyan parliament more transparent, as a lack of transparency has been considered a serious impediment to good governance in Kenya. Parliamentary business was essentially closed to the public, denying citizens the right to see, hear and assess what their elected representatives were doing until 2008 (USAID 2010b). HLB was conceived in 2008. New "House Rules" which were passed in April 2009, referred to as the Standing Orders, provided the legal basis for live broadcast. PSP introduced live audio and video to broadcast the proceedings from the main chamber and committee rooms in June 2009 with the funding from USAID/OTI (Office of Transition Initiatives) (SUNY/CID 2010; USAID 2010a). HLB has built structures and capacity within the Kenyan Parliament and it is believed that it would significantly increase the accountability of parliament to the citizens of Kenya, and boost public understanding of the parliament (USAID 2010b).

9.5.4 Parliaments Not Ready

Sometimes an argument is made that the parliament in the host country is not ready to accept and/or sustain the system. This argument can be made based on political unwillingness (e.g. see DAI 2008) and low capacities in the parliaments. One of the reasons for the latter is that many of the inductions in these institutions are made on political basis. Parliaments in developing countries face lack of institutional capacity, financial means (Hudson and Wren 2007) and political unwillingness to implement and sustain ICTs. They should be convinced from the information capabilities perspective about the benefits of supporting ICT initiatives. Telecasting can facilitate the members of the parliament in freedom of expression and dissemination of information. It allows the members to access the information about parliamentary activities. However, in present stage, telecasting usually does not facilitate the members in getting feedback from the citizens.

As mentioned earlier, the capability approach perspective tells us that the main emphasis of AAPS-projects should not be only to provide resources to the parliaments in the host country, but rather to design an environment, which will in realistic circumstances contribute to the information capabilities of citizens, members of parliaments and the parliaments. Stakeholders should take different initiatives such as identifying project champions, minimizing political unwillingness, and imparting training for parliamentary staff along with the parliamentary telecasting initiative as demonstrated by the following case.

Case 3 : Pakistan Legislative Strengthening Project (PLSP) was initiated in 2005 by the United States Agency for International Development (USAID) in collaboration with the Senate, the National Assembly and four provincial assemblies of Pakistan (DAI 2008). The project was implemented by Development Alternative Inc. (DAI) from 2005 to 2010. PLSP took many initiatives to strengthen the parliamentary institutions in Pakistan. As the resources and funding were available, PLSP took the telecasting initiative in combination of websites and web-casting systems development.[8] The main objective of the development of a dedicated parliamentary TV channel for telecasting the sessions of parliaments effort was to improve the representation function of the parliaments of Pakistan. PLSP faced many challenges including the resistance by the staff and the members of the parliaments. PLSP identified different project champions, lobbied, imparted training, and tried other tactics to convince the members and staff for the live telecasting. The initial plan was to install the telecasting equipment in the six houses. However, because of the resistance by members of parliaments and other challenges, telecasting equipment was setup in the National Assembly of Pakistan and one of the provincial parliaments.

9.6 Concluding Remarks

In this chapter, the trade-off between parliamentary telecasting initiative and Internet-based systems was analyzed from a capabilities approach. This approach can shed new light on decisions concerning the use and selection of technology. It was argued that during AAPS-projects, ICT systems should be designed in such a way that they actually contribute to expanding the capabilities of the citizens, members of parliaments, the parliaments, and other stakeholders. It was demonstrated that taking the conversion factors view can assist the stakeholders in choosing the appropriate technology. The capability approach points towards the importance of carefully choosing among technical alternatives in the given contextual environment, for which the relevant 'background conditions' should be analyzed. The information capabilities perspective further assists in the final choice. Parliamentary telecasting systems can create an environment of capability expansion for the parliamentary stakeholders. The chapter has shown, with the help of short illustrated cases, that some of the arguments for not investing in establishing the parliamentary telecasting systems lose force when seen through the lens of the capability approach. Thus the capability approach and information capabilities perspective can assist the decision makers in defending and investing in the parliamentary telecasting initiative. We should also remember that "it takes a robust, comprehensive technical infrastructure to support all of parliament's fundamental activities, but most importantly the political and institutional will to do so" (Griffith and Casini 2010, p. 20). In the end I would like to stress that

> When reaching out to a broader audience, we must be ready to concede that this will not necessarily lead to a better understanding of or greater public interest in politics. Confidence in politicians is not built solely on the debate in the chamber, it depends first and foremost on politics from a wider perspective, and also on the individual parliament. (Forsberg 2007)

---

[8] However, the systems were not utilized to their full potential.

Moreover, social arrangements may be decisively important in securing and expanding the freedom of the individual (Sen 1999). ICT systems – for example, parliamentary telecasting systems – can only act as the enabling factors and their success depends upon many other factors, including but not limited to the context of use, infrastructure, willingness of people, fitness with social, cultural, political, and legal values, norms, and practices.

Acknowledgement I am thankful to Jeroen van den Hoven, Marijn Janssen and Ilse Oosterlaken for their kind reviews, comments and suggestions.

References

Ahmed, M. A. (2008). *Developing parliamentary web portals for citizens, MPs and related groups – Challenges and proposed solutions* (pp. 1–7). IEEE International Symposium on Technology and Society.

Ahmed, M. A. (2010, July 12–15). *Improving information capabilities of parliamentary stakeholders through ICTs in developing countries.* Paper presented at the Worldcomp 2010 – The 2010 World Congress in Computer Science, Computer Engineering and Applied Computing, Las Vegas, Nevada, USA.

Ahmed, M. A. (2011a). Aid assisted parliamentary website initiatives – Challenges and solutions. In Z. Sobaci (Ed.), *E-parliament and ICT-based legislation: Concept experiences and lessons.* Hershey: IGI Global.

Ahmed, M. A. (2011b). ICTs for information capabilities of parliamentary stakeholders. In Z. Sobaci (Ed.), *E-parliament and ICT-based legislation: Concept experiences and lessons.* Hershey: IGI Global.

Alampay, E. A. (2006). Beyond access to ICTs: Measuring capabilities in the information society. *International Journal of Education and Development Using Information and Communication Technology, 2*(3), 4–22.

Alkire, S. (2002). Dimensions of human development. *World Development, 30*(2), 181–205.

Beetham, D. (2006). *Parliament and democracy in the twenty-first century: A guide to good practice.* Geneva: Inter-Parliamentary Union.

Dada, D. (2006). The failure of E-government in developing countries: A literature review. *The Electronic Journal of Information Systems in Developing Countries, 26*(7), 1–10.

DAI. (2008). *Pakistan legislative strengthening project quarterly report (October 1 – December 31, 2007).*

Fichtelius, E. (2007). The relationship between parliament, citizens and broadcasters is of growing importance. In IPU (Ed.), *The challenge of broadcasting parliamentary proceedings.* Geneva: IPU.

Forsberg, A. B. (2007). There is only one way forward for democracies: To work for greater openness and transparency. In IPU (Ed.), *The challenge of broadcasting parliamentary proceedings.* Geneva: IPU.

Garnham, N. (1999). Amartya Sen's "capability" approach to the evaluation of welfare: Its application to communications. In A. Calabrese & J.-C. Burgelman (Eds.), *Communication, citizenship and social policy: Rethinking the limits of welfare state* (pp. 113–123). Maryland: Rowman & Littlefield Publishers, Inc.

Gichoya, D. (2005). Factors affecting the successful implementation of ICT projects in government. *Electronic Journal of e-Government, 3*(4), 175–184.

Griffith, J., & Casini, G. (2010). *World e-Parliament Report 2010.* Global Centre for ICT in Parliament, Rome

Griffith, J., Griffith, J. B., & Casini, G. (2008). *World e-Parliament Report 2008. Global Centre for ICT in Parliament*.

Group, A. A. P. P. (2008). *Strengthening parliaments in Africa: Improving support*: World Bank Institute.

Heeks, R. (2002). Information systems and developing countries: Failure, success, and local improvisations. *The Information Society, 18*(2), 101–112.

Heeks, R. (2003). *Most EGovernment-for-development projects fail: How can risks be reduced?*. IDPM, University of Manchester, UK.

Hudson, A., & Wren, C. (2007). *Parliamentary strengthening in developing countries. Final report for DFID*. London: Overseas Development Institute.

IPU. (2007). *The challenge of broadcasting parliamentary proceedings*. Geneva: IPU.

IPU. (2009). *Guidelines for parliamentary websites – New edition*. Geneva: Inter-Parliamentary Union.

IWS. (2010). *Internet World Stats – Usage and population statistics*. Retrieved December 1, 2010, from http://www.internetworldstats.com/stats.htm.

Johnsson, A. B. (2007). The relationship between parliament, citizens and broadcasters is of growing importance. In IPU (Ed.), *The challenge of broadcasting parliamentary proceedings*. Geneva: IPU.

Johnstone, J. (2007). Technology as empowerment: A capability approach to computer ethics. *Ethics and Information Technology, 9*, 73–87.

Kleine, D. (2010a). The capability approach and the 'medium of choice': Steps towards conceptualising information and communication technologies for development. *Ethics and Information Technology, 13*(2).

Kleine, D. (2010b). ICT4WHAT?—Using the choice framework to operationalise the capability approach to development. *Journal of International Development, 22*(5), 674–692.

Leston-Bandeira, C. (2007). The impact of the internet on parliaments: A legislative studies framework. *Parliamentary Affairs, 60*(4), 665–674.

Madon, S. (2005). Evaluating the developmental impact of E-governance initiatives: An exploratory framework. *The Electronic Journal of Information Systems in Developing Countries, 20*(5), 1–13.

Mansell, R. (2001). *New media and the power of networks*. London: The London School of Economics and Political Science.

Musa, P., Mbarika, V., & Meso, P. (2006). *Integrating capability approach and cognitive constructivism to study technology acceptance in developing countries*. Paper presented at the Americas Conference on Information Systems, Acapulco, Mexico.

Nussbaum, M. (2000). *Women and human development: The capabilities approach*. Cambridge: Cambridge University Press.

Oosterlaken, I. (2009). Design for development: A capability approach. *Design Issues, 25*(4), 91–102.

Prakash, A., & De', R. (2007). Importance of development context in ICT4D projects: A study of computerization of land records in India. *Information Technology & People, 20*(3), 262–281.

Robeyns, I. (2003). Sen's capability approach and gender inequality: Selecting relevant capabilities. *Feminist Economics, 9*(2–3), 61–92.

Robeyns, I. (2006). The capability approach in practice. *Journal of Political Philosophy, 14*(3), 351–376.

Schacter, M. (2000). *Public sector reform in developing countries: Issues, lessons and future directions*. Ottawa: Policy Branch, Canadian International Development Agency.

SEAL. (2007). *QUARTERLY PROJECT REPORT [Q1, 2007] – United Nations Development Programme Afghanistan – Support to the Establishment of the Afghan Legislature (SEAL) – 01-04-2007–30-06-2007*. Afghanistan: UNDP.

Sen, A. (1980). Equality of what? In S. McMurrin (Ed.), *Tanner lectures on human values* (Vol. 1). Cambridge: Cambridge University Press.

Sen, A. (1985). Well-being, agency and freedom: The Dewey lectures 1984. *The Journal of Philosophy, 82*(4), 169–221.
Sen, A. (1999). *Development as freedom*. New York: Knopf.
Sen, A. (2002). *On ethics and economics*. Oxford: Blackwell Publishing.
SUNY/CID. (2010). Kenya Parliamentary Strengthening Program (PSP) 2000–2010. Retrieved Dec 1, 2010, from http://www.cid.suny.edu/our_work_projects_Kenya.cfm
UNDP. (2010). *UNDP | Democratic Governance*. Retrieved Mar 20, 2010, from http://www.undp.org/governance/sl-parliaments.htm
USAID. (2000). *USAID handbook on legislative strengthening* (pp. 20523–31000). Washington, DC: Center for Democracy and Governance Bureau for Global Programs, Field Support, and Research U.S. Agency for International Development.
USAID. (2010a). *Live broadcasts of parliamentary proceedings capture imagination of Kenyan public and change politics in Kenya*. Retrieved Dec 1, 2010, from http://kenya.usaid.gov/success-story/291
USAID (2010b). *Parliamentary strengthening program*. Retrieved Dec 1, 2010, from http://kenya.usaid.gov/programs/democracy-and-governance/570
Zheng, Y. (2007). *Exploring the value of the capability approach for E-development*. Paper presented at the 9th international conference on Social Implications of Computers in Developing Countries, Sao Paulo.
Zheng, Y., & Walsham, G. (2008). Inequality of what? Social exclusion in the e-society as capability deprivation. *Information Technology & People, 21*(3), 222–243.

Part III
Design

Chapter 10
Design, Risk and Capabilities

Colleen Murphy and Paolo Gardoni

10.1 Introduction

Risk is an inherent component of the design process and in particular of the engineering design process. As such, the design of complex artefacts, for example civil structural and infrastructural systems, is based on the premise that their performance can be predicted and evaluated with sufficient confidence for the engineer, client, and other stakeholders jointly to make intelligent and informed decisions. Such predictions are important to minimize disruptions to the operations of systems and contribute to sustainable designs. The uncertainties inherent in design give rise to risks. In the context of engineering design, risk refers more specifically to the probabilities of realizing different levels of performance of the designed artefact and the associated consequences. When designing an artefact, risk analysis can either serve as a final check for a possible design and be performed at the end of each design iteration, or be intrinsic in the design process, serving as a built-in constraint or objective function to minimize. The outcome of the risk analysis furthermore can affect the content of the design code, which in turn delimits the choices available to engineers.

Over the past few years there has been a significant improvement in the techniques for the assessment of the probabilities of potential consequences in risk analysis. However, there remains a crucial need for a comprehensive framework for conceptualizing and measuring the societal impact of the risks associated with

---

C. Murphy (✉)
Department of Philosophy, University of Illinois at Urbana - Champaign,
Urbana, IL, USA
e-mail: colleen@illinois.edu

P. Gardoni
Department of Civil Engineering, University of Illinois at Urbana - Champaign,
Urbana, IL, USA

selected designs. Moreover, recently specific engineering fields have adopted design codes rooted in a target probability, also known as reliability-based design codes. Such design codes assume a target probability of meeting certain specified performance levels and the code requirements are calibrated so that a design that follows the design requirements meets the targeted probability. However, reliability-based design codes only focus on probabilities and ignore the associated consequences and offer no general and logical way to select the target probability. Therefore, there is a need for a risk-based design that accounts in a normative and comprehensive way for the consequences associated to risks.

In past work, we have developed a capability approach to risk (Murphy and Gardoni 2006, 2007, 2008, 2010, 2011; Gardoni and Murphy 2008, 2009, 2010) and in particular to the quantification of the consequences associated to natural and man-made hazards. This chapter proposes a capability approach to design, herein called capability-based design. We argue that capabilities provide the requisite framework for conceptualizing consequences in risk analysis. In addition, a capability-based design code can properly account for both the probability and consequences dimensions of risk. Finally, a capability approach to design offers concrete guidance to engineers making choices in cases where an extensive design code does not exist and when balancing competing design constraints.

There are three sections in this chapter. The first provides a brief overview of the process of design in engineering. The second discusses the sources of risks in design and the role of the design codes. The third and final section articulates a capability-based design, which is premised on a capability approach to the determination, evaluation, and management of risk.

10.2 Engineering Design

Engineering design is a collaborative process. In addition to engineers, the design process can include technicians, architects, designers, managers and other stakeholders, including owners. Engineering design problems are characteristically ill-structured (van de Poel 2001), or to use terminology from mathematics ill-conditioned. The range of alternative solutions to a given design problem is not well-specified. That is, it is not possible to make a complete list of potential solutions. In addition, there is no given criterion for what constitutes a good solution. Thus it is not possible to rank alternative options as better or worse solutions. Part of the reason that there is more than one solution and difficulty ranking solutions is that not all of the inputs in the design process are well-defined a priori. One consequence of the ill-structured nature of design problems is that the design process is often iterative.

Engineering design problems become more structured through the introduction of external constraints. Such constraints are used to specify the criteria a design should satisfy and to set boundaries for the solution space. Constraints can include both normative and technical considerations. Examples of external constraints include dimensional constraints, safety norms through a target probability of failure, sustainability guidelines, and cost understood either in terms of construction costs or life-cycle costs. Life-cycle costs include construction costs as well as maintenance and repair costs over the life of a structure. In cases of normal design, there are standard strategies to operationalize such constraints and typical code requirements that help delimit and guide the design process. For example, a sustainability criterion for cars can be satisfied by designing a lightweight car. Passenger protection mechanisms like airbags are standardly used to satisfy safety requirements (van de Poel and van Gorp 2006). By contrast, in cases of radical design, the way in which a given external constraint should be operationalized may be unknown or subject to revision. Radical design is intrinsically more novel and exploratory in nature.

Once the solution space and criteria for success are articulated, design problems are often divided into smaller problems, with alternative solutions to these smaller problems devised. For example, for many modern products that contain parts or sub-systems, each part or sub-system is designed separately. Parts that are lower in the design hierarchy, insofar as they comprise a smaller role in the overall product, must be designed while respecting the external constraints that the system as a whole and/or sub-systems into which a given part fits need to satisfy (van de Poel and Gorp 2006).

The final step in the design process is to select the solution that will be pursued from among a range of alternative options. The chosen design strikes the best compromise in meeting design requirements. The chosen alternative will form the foundation for the artefact that is ultimately produced through the production process.

10.3 Risk

Risk is an inherent part of the design process. In this section we discuss the sources of uncertainty and risk in design, with a particular focus on the design of civil structures and infrastructure.

The design of complex artefacts, for example civil structural and infrastructural systems, is based on the premise that their performance can be predicted and evaluated with sufficient confidence for the engineers, clients and other stakeholders jointly to make intelligent and informed decisions. Such predictions are important to minimize disruptions to the operations of systems and contribute to sustainable designs. Both the minimization of operational disruption of systems and the optimal decision making for sustainable system management necessitate a shift away from current prescriptive codes, which tend to be implicitly conservative and do not properly account for the consequences of damage or failure of an artefact, and toward a design process and design codes more firmly rooted in the realistic prediction of loads, structural behavior, probabilities of damage and failure, and the associated consequences.

Models play an important role in the process of predicting the performance of structural and infrastructural systems. Through models, engineers can examine how an object will perform given certain conditions. However, there are a number of uncertainties surrounding the process of modeling, which limit the ability of engineers to predict with compete accuracy how a given product will perform (Murphy et al. 2011). Uncertainties arise because of the approximations and assumptions used to develop a model. A model may be inexact because, for example, variables are not included which should be included because they influence the performance. A model may assume normality or homoskedasticity, when these assumptions are violated. Furthermore, models may be calibrated based on too sparse a sample size of data, giving rise to statistical uncertainty, or on errors in data due to errors in the measurement process for the data. In applying or implementing a model, uncertainties may arise because of the possibility of human error in incorrectly using a model in an inapplicable area or because of variability in the basic variables that are inputs in a model (e.g., variability in the characteristics of actual steel used in a bridge) (Murphy et al. 2011). Moreover, there are limits in the lab tests and simulations used to test a model; such tests cannot duplicate every possible situation that may arise in the world, both because of time constraints and because of limits on the ability to predict the possible situations that may be relevant for assessing a product's performance (Wetmore 2008).

The uncertainties inherent in engineering design give rise to risks. *Risk* typically refers to the *probabilities* of potential *consequences* in various hazard scenarios (Vose 2000; Bedford and Cooke 2001). In the context of design, risk refers more specifically to the probabilities of different levels of performance of the designed artefact and the associated consequences, which can be, as Murphy and Gardoni (2006) noted, both positive (benefits) and negative (losses). Examples of possible performance levels for a structure under a specific load include collapse, severe damage, minor damage, and no damage. Positive consequences include both the benefits to the society brought by the proper functioning of the artefact (no damage state), and the opportunities brought by the other damage states.

Risk analysis is the process of determining, evaluating and managing risk. *Risk determination* refers to the quantification of the probabilities of different levels of performance of the designed artefact and the associated consequences. Once one has determined the risks associated with a given design (or design code), the next step is to evaluate the risks. *Risk evaluation* refers to the assessment of the outcomes of the risk determination. In particular a risk can be deemed acceptable or not. Such evaluation typically involves engineers but also takes into consideration the preferences of the general public. In risk evaluation different kinds of judgments may be reached. It may be determined that risks can and should be reduced through minor improvements in the design (or design code). Other risks may require more significant revisions. Yet other risks may be such that a given solution should be abandoned. The outcomes of risk evaluation provide the inputs for risk management. *Risk management* refers to the strategies and specific actions that individual engineers, regulatory agencies, and engineering professional societies undertake to deal with risks. Such strategies and actions should account for typically limited resources and time and include, for example, changes in the design specifications and in the details of actual design. More far-sighted management action goes beyond a specific project, including suggested changes in the design code requirements and involving specific supervising engineering committees.

In design, a probabilistic approach that properly accounts for the underlying uncertainties is essential for the identification and quantification of performance parameters, for assessing the likelihood of meeting selected performance levels, for assessing the reliability of structural components and systems, for addressing life-cycle cost and sustainability issues, and for an overall consistent treatment of risk. When designing an artefact, risk analysis can either serve as a final check for a possible design and be performed at the end of each design iteration, or be intrinsic in the design process serving as a built-in constraint or objective function to minimize. In cases of normal design (van de Poel and van Gorp 2006), a detailed design code typically exists, and so the outcomes of the risk analysis reflect the risks associated both with the engineers' choices and with the choices prescribed by the design code. In cases of radical design (van de Poel and van Gorp 2006), a detailed code might not be available and so the outcome of a risk analysis reflects primarily an engineer's choices. The outcome of the risk analysis furthermore can affect the content of the design code, which in turn delimits the choices available to engineers.

10.4 Capabilities

In this section we discuss how the capability approach can contribute to engineering design. We first provide a brief overview of the idea of capability. We then show how the capability approach can be used in engineering design. Finally we discuss the advantages of using a capability approach, including how using a capability approach can improve the outcome of the design.

The capability approach offers a distinctive space for making evaluations about states of affairs, including the well-being of individuals and the development of communities. This approach is currently used for evaluations in fields as varied as social choice theory, development economics, and disaster studies (Robeyns 2006; Gardoni and Murphy 2010).

In the capability approach, evaluations of states of affairs are made by considering the genuine opportunity or effective freedom that individuals enjoy to achieve valuable states and activities, or functionings (Sen 1989, 1992, 1993, 1999a, b; Nussbaum 2000a, b, 2001). Examples of functionings include being adequately nourished, educated, and mobile. An individual has a given capability a when she has a genuine opportunity to achieve a given functioning a. That is, an individual has a capability to be educated if she enjoys a genuine opportunity to be educated.

Genuine opportunities, or capabilities, are influenced by three general considerations: what an individual has, what she is able to do what she has, and the security with which an individual is able to achieve a given functioning (Wolff and de-Shalit 2007). What an individual has broadly captures the internal and external resources at her disposal. Internal resources include talents and skills.

External resources include income and family support. What an individual is able to achieve with her resources depends on the social and material structure in which she operates (Sen 1993, 1999b). This structure encompasses factors such as laws, social norms, and the built physical infrastructure of a community. Mobility, for example, is influenced by the roads and systems of transportation available in a community. Poorer roads make transportation more difficult. However, mobility is also influenced by the resources an individual possesses, such as income in a community where public transportation is the norm or a vehicle in communities where public modes of transportation are limited. Moreover, public or private modes of transportation will enhance mobility only if individuals are not legally or socially barred from using them.

In practice capabilities are interdependent. That is, the freedom an individual has to achieve a given functioning will be influenced by which other functionings she chooses to achieve. For individuals with little income in a community in which housing and food are available only through monetary purchase and little government subsidy or assistance is available, resource constraints might mean they are able to be housed or adequately nourished, but not both at the same time. In light of this interdependence, we can speak of an individual's general capability, which refers to her freedom to achieve different vectors of functionings (Sen 1992).

Recognition of the fact that it is only vectors of capabilities that are open to individuals draws attention to the need to consider the security with which a given functioning can be achieved (Wolff and de-Shalit 2007). An individual might be able to achieve a given functioning, but only by putting other functioning achievements at considerable risk. A functioning is at considerable risk when it is able to be achieved only temporarily or by involuntarily putting other achievements at risk. When the only forms of employment are dangerous occupations, then an individual may be employed but only by putting her bodily integrity at risk (Wolff and de-Shalit 2007).

The capability approach has a distinctive perspective on how to conceptualize the well-being of individuals. Well-being can refer to either the advancement of an individual's own welfare or the advancement of her agency goals, which captures the broader category of objectives an individual has reason to pursue (Sen 2009). Capability can capture either an individual's welfare freedom or her agency freedom. Greater capability understood as agency freedom might not always lead to greater capability understood as individual welfare freedom. Some goals that an individual chooses to pursue might be in tension with her individual welfare.

The distinctiveness of the evaluative space of capabilities for understanding well-being is clear when we consider two other evaluative spaces, resources and utilities. The resource framework evaluates well-being based on the market commodities (e.g., income) or primary goods (e.g., rights, income, and the social bases of self-respect) that an individual possesses. Limitations of conceptualizing well-being in this way include the following. First, this framework focuses on the means for well-being instead of well-being itself. Income is valued for what it allows us to do and become, not for its own sake. Second, this framework does not take into consideration variability in what individuals have an opportunity to do with the same amount of resources, given the social and material environment in which they act. Two individuals might both possess a bike, but only one may be able to use it. Social norms and a poor built infrastructure limit the ability of the other individual to use the bike for mobility.

Alternately, well-being can be conceptualized in terms of utility, where utility is understood to capture happiness defined as desire-fulfillment (Sen 2009). This framework does not focus on means to well-being but on well-being itself; individuals value happiness for its own sake. However, well-being defined as utility has a fundamental limitation: happiness does not capture all of well-being. Individuals adjust their desires to their circumstances, and so an individual who is objectively deprived may desire very little and have those few desires satisfied. From a utility approach she would be judged happy. Conversely, an individual who has objectively quite a bit may develop desires that are difficult to satisfy and so have fewer desires satisfied. This second individual would be judged less happy and so enjoying less well-being than the first (Crocker 2008). As these two cases illustrate, factors other than individual happiness must be considered to get a comprehensive picture of the well-being of individuals. The capability approach does precisely this. It can incorporate the capability to be happy as one capability, while acknowledging that well-being involves a greater range of opportunities that it is important for individuals to enjoy.

10.5 A Capability Approach in Design

In this section we first discuss how the capability approach can inform how risk is accounted for in design. Our discussion considers the capability approach to risk determination, risk evaluation, and risk management. We then highlight the benefits of a capability approach to design.

10.5.1 Risk Determination

In the process of determining risks, the capability approach can provide the basis for conceptualizing and assessing the consequence component of risk. In a capability approach, the consequences of hazardous scenarios are conceptualized and assessed in terms of changes in capabilities. Risk is then defined as the probability that capabilities will be reduced (Murphy and Gardoni 2006). The consequence component of risk can be captured using the general framework that Gardoni and Murphy (2009) developed first in the context of natural hazards. In particular, a Hazard Impact Index (*HII*) can be formulated, which echoes the formulation of the Human

Development Index (*HDI*) already available. The following four steps are needed to compute the *HII*:

1. Select capabilities that are relevant for the risk under consideration and can capture the associated consequences
2. Select indicators for the chosen capabilities
3. Convert each indicator into an index that ranges from 0 (minimum achievement) and 1 (maximum achievement)
4. Combine all indices through an averaging process to create an aggregate measure

To assist in the selection in Step 1, Gardoni and Murphy (2009) offered a list of criteria that capabilities should meet. In particular, Gardoni and Murphy (2009) argue that a capability is relevant when it captures the values and concerns inherent with the risk under consideration. Furthermore, for a successful practical implementation of a capability approach to risk determination, the most parsimonious number of capabilities should be selected. Parsimony requires that each considered capability brings in unique information. Limiting the duplication of information is also important not to double count any particular dimension of well-being. Gardoni and Murphy (2009) also proposed a list of capabilities for risk determination. While Gardoni and Murphy (2009) worked specifically in the context of natural hazards, the criteria and list that they provide is more generally applicable to risk determination in the context of design. In addition, natural hazards are a subset of the possible loading scenarios that an artefact needs to be designed to sustain.

An indicator is a measurable quantity that follows the impact on the corresponding capability, which is not directly measurable (Gardoni and Murphy 2010). Indicators are used in Step 2 to measure the relative change in the level of achievement of each associated functioning (Gardoni and Murphy 2010).

In Step 3 indices are created through a scaling process analogous to that in the formulation of the *HDI*. The indices are dimensionless and represent each indicator on a scale from 0 to 1 so that the levels of achievement of each selected capability can be compared and combined as described in Step 4.

As noted earlier, capabilities are not independent; the achievement of one functioning can have implications for the freedom of an individual to achieve other functionings. Thus, it is necessary to determine the vectors of capabilities that are open to individuals. Moreover, one limitation with the *HDI* is that it actually measures the achieved functionings instead of capabilities (Murphy and Gardoni 2010). To capture vectors of capabilities Murphy and Gardoni (2010) proposed the following method, which is based on the idea that the functionings achievements of individuals similarly situated provides information about the possibilities open to a given individual. That is, information about individuals similarly situated helps us distinguish between a case in which an opportunity is not open to an individual from a case in which an opportunity is open to an individual, though she declines to pursue it. To compute vectors of capabilities, the overall impact of a hazard on each individual is computed first. This provides information about changes in achieved functionings for each individual across a community. Once this information is converted into an index, we compute the average and standard deviation of the impact of a hazard across a defined set of individuals. The standard deviation captures the range of variation in the impact across individuals in a community.

The same framework can be used both for risk and for disaster evaluation. Disaster evaluation measures the actual impact of a hazardous scenario that occurred. In risk determination, a hazard has the potential to occur and the values of the indicators need to be predicted (Gardoni and Murphy 2010).

A capability approach is not needed for the assessment of the probability component of risk. Several advanced techniques are already available to assess probabilities, including analytical or numerical integration, simulation, moment-based methods, or first- and second-order methods (FORM/SORM) (Ditlevsen and Madsen 1996; Gardoni et al. 2002). The advent of new technologies that enable engineers to collect, manipulate, and display extensive data with minimal human intervention (including sensor and information technologies) is also drastically increasing the diffusion and implementation of probabilistic methods and statistics, helping to reduce the statistical uncertainty associated to the limited number of data, and promoting a more sustainable and resilient built environment (Huang et al. 2009).

10.5.2 Risk Evaluation

Once one has determined the risks associated with a given design, the next step is to evaluate the risks. How should we judge whether risks are acceptable? In Murphy and Gardoni (2008), we argue that a capability approach offers a basis for such evaluation. The key idea behind the capability approach to acceptable risk we propose is that as a matter of justice individuals should enjoy, and communities should be prepared to guarantee, a certain minimum threshold level of capability (Nussbaum 2000b). However, the world is not deterministic; individuals and communities do not know with certainty what will happen in the future, exactly how a designed artefact will behave, or whether their efforts to guard against certain hazards will be successful. Thus, this demand of justice must be understood probabilistically. Justice requires that it be sufficiently likely that levels of capabilities will be maintained at or above the specified minimum threshold level. When thinking explicitly about the acceptability of risk from a capability approach, a risk is acceptable, then, insofar as a certain threshold level of capability is sufficiently likely to be maintained (Murphy and Gardoni 2008). A framework for acceptable risk specifies what the threshold level of capability is and the requisite probability that individuals will not fall below this threshold. The process of specification can be informed by or the product of public deliberation. For the threshold to capture a realistic ideal, it must reflect what it is feasible for a community to guarantee individuals given scarce resources.

10.5.3 Risk Management

What are the implications of the evaluation of a risk as acceptable or not for design? We answer this question by looking at the implications of a risk's acceptability for what an engineer should do, either as a member of the engineering community determining the design code or as an individual engineer designing a particular artefact within the constraints set by the design code.

Turning first to the design code, current design codes are typically prescriptive primarily rooted in what "worked in the past." Only recently reliability-based design codes have been developed in specific engineering fields. Such design codes assume a target probability of failure and define code requirements so that ultimately a designed artefact has a probability of failure close to the target one. One fundamental limitation with reliability-based design is that it only accounts for the probability of failure of a given design, but ignores the associated consequences. Therefore, a risk-based design is preferable because it captures both the probability and the associated consequences. When a capability approach is used in the risk determination of a risk-based design, then we can talk about *capability-based design*.

A risk-based design code should have a target risk instead of a target probability that each design needs to satisfy. The specific requirements of the design code are then developed so that a design that follows them will meet the target risk. Specifically, for a capability-based design, the threshold of acceptable risk provides the target risk. Moreover, the threshold of acceptable risk can guide the formulation of the particular requirements laid out in the design code. For example, the process of formulating the specific requirements for a bridge, such as the minimum amount of steel that is needed and the minimum cover for the steel, would be guided by what an overall set of minimum requirements would need to be like for a given structure to meet or exceed the threshold of acceptable risk.

Engineers will design specific artefacts within the constraints set by the design code. However, the capability approach can provide guidance for particular design decisions with which engineers will be faced. First, there may be designs that fail to satisfy the threshold of acceptable risk. This is especially likely in cases of radical design, where there does not exist a particular code to guide the design process. In such cases, an engineer must modify a particular design so that the overall risk meets the minimum risk spelled out in the acceptable threshold. The capability approach can provide additional guidance on what kind of modification to a given design should be pursued (Murphy and Gardoni 2007). In particular, engineers can look for the modification that is the most efficient, in terms of a dollar per unit reduction in the overall risk.

In other instances, however, an engineer may meet the particular requirements laid out in the code and also have an overall design that satisfies the acceptable risk threshold. Here an engineer may be faced with the question of how to balance a concern with protecting capabilities against the additional external constraints that a given design should satisfy, such as sustainability or aesthetic considerations. A design need not minimize the risk to capabilities. Minimizing risk to capabilities may both stifle innovation generally and also prevent an engineer from giving weight to other external constraints, which in many instances represent important ethical values. The capability approach does not provide a comprehensive framework for answering what kind of trade-off or balance of competing external constraints is best in each case. In some cases, external constraints could also be captured using a capability approach insofar they affect individual well-being (e.g., Gardoni and Murphy 2008). Then judgments about how to compare and weigh different dimensions of well-being would need to be made. In other cases, comparisons among different kinds of goods are necessary. Here the capability framework provides a reminder of what is being increased, namely the possibility of harming well-being, when risks associated with a particular design are increased, or the possibility of promoting well-being, when a design is successful.

10.6 Benefits of a Capability-Based Design

As noted earlier, over the past few years there has been a significant improvement in the techniques for the assessment of the probabilities of potential consequences. There is also a richer understanding of the empirical trends in the societal vulnerability and recovery process, motivated in part by a growing recognition by engineers that a proper and complete account of the *consequences* that captures the complex societal impact needs to incorporate the work from social science (An et al. 2004). As An et al. (2004) write, "Over the past decade, earthquake engineering and similar natural hazard based research activities have begun to integrate social science questions into technical research agenda." They also argue that this need, at least to some extent, led to the creation of three National Science Foundation (NSF) Engineering Research Centers: the Multidisciplinary Center for Earthquake Engineering (MCEER), the Pacific Earthquake Engineering Research Center (PEER), and the Mid-America Earthquake Center (MAE Center). Part of the outcome of the research conducted at these centers is an extensive literature in sociology and disaster studies that is now available that focuses in particular on case studies on the effects of past disasters on households and vulnerable segments of the population (e.g., Van Willigen et al. 2005; Kajitani et al. 2005; Dash and Gladwin 2007; Dash et al. 2007). Such research often identifies social reasons for the different rates of recovery among families, individuals, communities, or businesses (see, for example, Bates and Peacock 1992; Tierney 1992, 1994; Nigg 1996; Peacock and Girard 1997; Dahlhamer and D'Souza 1997; Mileti 1999; Fothergill et al. 1999; Petterson 1999).

However, there remains a crucial need for a comprehensive framework not for explaining why there was a given the impact, but for conceptualizing and measuring the societal impact of the risks associated with selected designs itself. The capability approach is well-suited to provide the needed framework, in part because it has certain additional advantages over two common ways of conceptualizing the consequences, the resource and utilitarian frameworks.

Traditionally, engineers and social scientists have assessed consequences from a resource-based framework (Rowe 1980; Vose 2000; Bedford and Cooke 2001; Haimes 2004). In this framework, consequences are defined in terms of largely market resources lost, including time (e.g., through delays in construction), money, structures, and individuals. The general problem with this framework is that information about lost resources does not tell us how the lives of individuals or communities are affected. There is variation in what individuals and communities can do with a given set of resources because of variation in the social and material structure within which individuals act. The loss of one structure may be devastating for individuals within a community if that structure is the only hospital in a large area or an important historical holy site for members of a particular religion. In such cases, the ability of individuals to maintain their health when ill through the provision of medical assistance or the ability of individuals to practice a given religion may be undermined. However, the loss of one structure may have little impact on a community if that community has the resources to rebuild it if desired and the structure itself housed a small commercial enterprise with little bearing on a community. By contrast, the interest of the capability approach is directly on how the opportunities of individuals are affected, and takes into account both lost resources and the loss of dimensions of the social and material environment that are needed to convert resources into opportunities. That is, the capability approach considers how a hazard will change the opportunities of individuals.

A second common framework for risk analysis conceptualizes consequences in terms of utility. Utility is understood to reflect happiness, where happiness is a function of preference satisfaction. Preference satisfaction is in turn defined in monetary terms, based on the willingness of an individual to pay to avoid being exposed to certain risks (Sunstein 2005). In practice, market activity or contingent valuation surveys are used to assess the willingness to pay (Sunstein 2005; Hansson 2007). The utility framework shares the same limitation as the resource framework: it does not provide an accurate assessment of the impact of hazards, because it either over- or under-estimates their importance (Slovic 1987; Murphy and Gardoni 2006). Utility estimations based on market activity are often inaccurate first because individuals may not be fully knowledgeable about the risks they face when making certain choices about, for example, types of employment to pursue. Thus, they make seem willing to accept risks at a much lower cost than they would if fully informed. Moreover, even when fully informed, individuals may not be free to accept or reject certain risks solely on the basis of their valuation of certain costs (Anderson 1988; Wolff and de-Shalit 2007). Individuals may accept certain risks associated with certain housing because that is the only type of housing that is available and affordable. By contrast, the capability approach considers the loss of genuine opportunities associated with a hazardous scenario.

In addition to providing a more accurate characterization of risk, capability-based design has benefits both for the design code and the design process. As we discussed in the previous section, currently design codes are typically based on what "worked in the past." As such, it is not known how safe a design that follows the design code is, because the design code is not based on a target probability but on a set of rules that contingently worked in the past. This is detrimental because

1. designs tend to be too conservative and artefacts overdesigned, which leads to a waste of limited resources;
2. there is no consistent reliability among artefacts (some are much safer than others);
3. any attempt not to overdesign an artefact might lead to an unsafe solution.

Only recently have specific engineering fields put forward and adopted design codes rooted in a target probability, also known as reliability-based design codes. While such design codes clearly assume a target probability of meeting certain specified performance levels and the code requirements are calibrated so that a design that follows the design requirements meets the targeted probability, reliability-based design codes still have two limitations:

1. they only focus on probabilities and ignore the associated consequences;
2. there is no general and logical way to select the target probability.

Increasingly, engineers recognize these limitations. In practice, target probabilities are typically inversely proportional to their consequences. That is, the greater the consequences of failing to meet a specified performance level, the lower the target probability must be. The specific target probability for a given artefact can be selected so that the corresponding risk is comparable in magnitude to those associated to other artefacts. Consequences, as noted above, are typically measured in terms of fatalities or monetary terms. However, this *de facto* comparative risk-based approach has two main limitations. First, the target probability is selected assuming a typical value for the consequences. However modifications in a design can both reduce the probability and the consequences. Thus, a risk-based design code should target both probabilities and consequences. Second, a monetary formulation of the consequences has the limitations discussed above. This clearly call for a capability-based design code that takes into consideration both probability and consequences and captures consequences using a capability approach.

Finally, a capability-based design has advantages for engineers designing specific artefacts. A capability-based design provides the most valuable and complete guidance to an engineer in the design process. An engineer designing an artefact can check that the designed artefact meets all the code requirements and, in particular, the target acceptable risk. In radical design, a target risk can be a valuable (perhaps one of only a few) guide for an engineer. Furthermore, a capability-based design can provide some guidance to engineers as they make tradeoffs between risk and meeting other design constraints, some of which may be also translated in terms of capabilities. Moreover, engineers can explore possible modifications to the design that are most effective in reducing risk with the smallest compromise in the benefits to well-being provided by the designed artefact in the event there is no failure. Finally, a capability-based design has a central principled advantage, namely, it puts the well-being of individuals as a central focus of the design process.

Acknowledgements This research was supported primarily by the Science, Technology, and Society Program of the National Science Foundation Grant (STS 0926025). Opinions and findings presented are those of the authors and do not necessarily reflect the views of the sponsor.

References

An, D., Gordon, P., Moore, J. E., II, & Richardson, H. W. (2004). Regional economic models for performance based earthquake engineering. *Natural Hazards Review, 5*(4), 188–195.

Anderson, E. (1988). Values, risks, and market norms. *Philosophy and Public Affairs, 17*(1), 54–65.

Bates, F., & Peacock, W. G. (1992). Measuring disaster impact on household living conditions. *International Journal of Mass Emergencies and Disasters, 10*(1), 133–160.

Bedford, T., & Cooke, R. (2001). *Probabilistic risk analysis: Foundations and methods*. Cambridge: Cambridge University Press.

Crocker, D. (2008). *The ethics of global development*. New York: Cambridge University Press.

Dahlhamer, J., & D'Souza, M. (1997). Determinants of business disaster preparedness. *International Journal of Mass Emergency and Disasters, 15*, 265–281.

Dash, N., & Gladwin, H. (2007). Evacuation decision making and behavioral responses: Individual and household. *Natural Hazards Review, 8*(3), 69–77.

Dash, N., Morrow, B. H., Mainster, J., & Cunningham, L. (2007). Lasting effects of Hurricane Andrew on a working-class community. *Natural Hazards Review, 8*(1), 13–21.

Ditlevsen, O., & Madsen, H. O. (1996). *Structural reliability methods*. New York: Wiley.

Fothergill, A., Maestas, E. G. M., & Darlington, J. D. (1999). Race, ethnicity and disasters in the United States: A review of the literature. *Disasters, 23*(2), 156–173.

Gardoni, P., & Murphy, C. (2008). Recovery from natural and man-made disasters as capabilities restoration and enhancement. *International Journal of Sustainable Development and Planning, 3*(4), 1–17.

Gardoni, P., & Murphy, C. (2009). A capabilities-based approach to measuring the societal impacts of natural and man-made hazards. *Natural Hazard Review, 10*(2), 23–37.

Gardoni, P., & Murphy, C. (2010). Gauging the societal impacts of natural disasters using a capabilities-based approach. *Disasters: The Journal of Disaster Studies, Policy and Management, 34*(3), 619–636.

Gardoni, P., Der Kiureghian, A., & Mosalam, K. M. (2002). Probabilistic capacity models and fragility estimates for RC columns based on experimental observations. *ASCE Journal of Engineering Mechanics, 128*(10), 1024–1038.

Haimes, Y. Y. (2004). *Risk modeling, assessment, and management*. Hoboken: Wiley Series in Systems Engineering and Management.

Hansson, S. O. (2007). Philosophical problems in cost-benefit analysis. *Economics and Philosophy, 23*, 163–183.

Huang, Q., Gardoni, P., & Hurlebaus, S. (2009). Probabilistic capacity models and fragility estimates for reinforced concrete columns incorporating NDT data. *ASCE Journal of Engineering Mechanics, 135*(12), 1384–1392.

Kajitani, Y., Okada, N., & Tatano, H. (2005). Measuring quality of human community life by spatial-temporal age group distributions: case study of recovery process in a disaster-affected region. *Natural Hazards Review, 6*(1), 41–47.

Mileti, D. S. (1999). *Disasters by design: A reassessment of natural hazards in the United States*. Washington, DC: Joseph Henry Press.

Murphy, C., & Gardoni, P. (2006). The role of society in engineering risk analysis: A capabilities-based approach. *Risk Analysis, 26*(4), 1085–1095.

Murphy, C., & Gardoni, P. (2007). Determining public policy and resource allocation priorities for mitigating natural hazards: A capabilities-based approach. *Science and Engineering Ethics, 13*(4), 489–504.

Murphy, C., & Gardoni, P. (2008). The acceptability and the tolerability of risks: A capabilities-based approach. *Science and Engineering Ethics, 14*(1), 77–92.

Murphy, C., & Gardoni, P. (2010). Assessing capability instead of achieved functionings in risk analysis. *Journal of Risk Research, 13*(2), 137–147.

Murphy, C., & Gardoni, P. (2011). The capability approach in risk analysis. In S. Roeser (Ed.), *Handbook on risk theory* (pp. 979–997). Dordrecht: Springer.

Murphy, C., Gardoni, P., & Harris, C. E. (2011). Classification and moral evaluation of uncertainties in engineering modeling. *Science and Engineering Ethics, 17*(3), 553–570.

Nigg, J. (1996). Anticipated business disruption effects due to earthquake-induced lifeline interruption. In F. Cheng & Y. Wang (Eds.), *Post-earthquake rehabilitation and reconstruction* (pp. 47–57). St. Louis: Pergamon.

Nussbaum, M. (2000a). Aristotle, politics, and human capabilities: A response to Antony, Arneson, Charlesworth, and Mulgan. *Ethics, 111*(1), 102–140.

Nussbaum, M. (2000b). *Woman and human development: The capabilities approach*. Cambridge: Cambridge University Press.

Nussbaum, M. (2001). Adaptive preferences and women's options. *Economics and Philosophy, 17*, 67–88.

Peacock, W. G., & Girard, C. (1997). Ethnic and racial inequalities in hurricane damage and insurance settlements. In W. G. Peacock, B. H. Morrow, & H. Gladwin (Eds.), *Hurricane Andrew: Ethnicity, gender and the sociology of disasters*. London: Routledge.

Petterson, J. (1999). *A review of the literature and programs on local recovery from disaster* (Working Paper #102). Natural Hazards Research and Applications Information Center, Institute of Behavioral Science, University of Colorado. Published by Public Entity Risk Institute. www.riskinstitute.org. Accessed 3 Dec 2007.

Robeyns, I. (2006). The capability approach in practice. *Journal of Political Philosophy, 14*(3), 351–376.

Rowe, W. D. (1980). Risk assessment: Theoretical approaches and methodological problems. In J. Conrad (Ed.), *Society, technology, and risk assessment* (pp. 3–29). New York: Academic Press.

Sen, A. (1989). Development as capabilities expansion. *Journal of Development Planning, 19*, 41–58.

Sen, A. (1992). *Inequality reexamined*. Cambridge: Harvard University Press.

Sen, A. (1993). Capability and well-being. In M. Nussbaum & A. Sen (Eds.), *The quality of life* (pp. 30–53). Oxford: Clarendon.

Sen, A. (1999a). *Commodities and capabilities*. Oxford: Oxford University Press.

Sen, A. (1999b). *Development as freedom*. New York: Anchor Books.

Sen, A. (2009). *The idea of justice*. Cambridge: Belknap Press of Harvard University Press.

Slovic, P. (1987). Perception of risk. *Science, 236*, 280–285.

Sunstein, C. (2005). Cost-benefit analysis and the environment. *Ethics, 115*, 251–285.

Tierney, K. (1992). What are the likely categories of loss and damage? In *The economic consequences of a catastrophic earthquake: Proceedings of a forum* (pp. 77–82). Washington, DC: National Academy Press.

Tierney, K. (1994). Societal impacts. In J. Goltz (Ed.), *The Northridge, California earthquake of January 17, 1994: General reconnaissance report* (pp. 7:1–7:10). Buffalo: National Center for Earthquake Engineering Research.

van de Poel, I. (2001). Investigating ethical issues in engineering design. *Science and Engineering Ethics, 7*, 429–446.

van de Poel, I., & van Gorp, A. C. (2006). The need for ethical reflection in engineering design: The relevance of type of design and design hierarchy. *Science, Technology & Human Values, 31*(3), 333–360.

Van Willigen, M., Edwards, E., Lormand, S., & Wilson, K. (2005). Comparative assessment of impacts and recovery from Hurricane Floyd among student and community households. *Natural Hazards Review, 6*(4), 180–190.

Vose, D. (2000). *Risk analysis: A quantitative guide.* New York: Wiley.

Wetmore, J. M. (2008). Engineering with uncertainty: Monitoring air bag performance. *Science and Engineering Ethics, 14*, 201–218.

Wolff, J., & de-Shalit, A. (2007). *Disadvantage.* Oxford: Oxford University Press.

Chapter 11
Re-conceptualizing Design Through the Capability Approach

Crighton Nichols and Andy Dong

11.1 Introduction

The appropriate technology movement has been steadily building since Schumacher popularized the idea in his seminal work *Small is Beautiful* (1973). The set of technologies developed under the banner of 'appropriate' is inspiring to say the least, and has made significant inroads into solving basic development problems, such as access to clean drinking water, adequate sanitation, and hygienic food preparation.

At first glance, the idea of designing a technology that is sensitive to the cultural, political and economic conditions of a community sounds appealing. How can we argue that introducing technology based on locally available materials to provide clean drinking water is 'wrongheaded'? Some members of the design for development community, though, have been critically reflecting on the efforts of product designers to improve the lives of impoverished and marginalized people through the introduction of appropriate technology. For example, David Stairs, in critiquing the Cooper-Hewitt National Design Museum exhibition *Design for the Other 90%*, questioned the extent to which the exhibited design interventions responded to the lived exigencies of the world's poor (Stairs 2007). The disconnect between designers and the communities they claim to serve, in Stairs' critique, is representative of one dimension of this critical reflection, which is about whether appropriate technologies are 'appropriate' and the criteria for 'appropriate'. Who decides if a

---

C. Nichols (✉) • A. Dong
Design Lab, Faculty of Architecture, Design and Planning, University of Sydney,
Wilkinson Building G04, Sydney NSW 2006, Australia
e-mail: crighton.nichols@gmail.com; andy.dong@sydney.edu.au

technology is appropriate and under which considerations? What happens when criteria are in conflict?[1]

Discussions about the ethical consequences of technology and the values that their developers implicitly promote through the introduction of the technology are important to have, because they are ensconced in concerns for social justice, such as in eliminating economic and social disparities. Adjudicating 'what is appropriate' can help us to answer questions surrounding the role of technology in social justice, and if the *design* of the technology has any bearing on social justice. These debates help us decide which design for development programs take us closer to social justice, and which technologies contribute to equality of opportunity, and what outcomes we should assess.

At the same time, we are perhaps overly self-critical of our achievements in appropriate technology when the introduction of every technology has positive, negative and unintended consequences. The introduction of every technology presents a moral problem (e.g., Should parents allow children to use social media? Should we introduce laptops to remote Indigenous Australian communities?). The introduction of every technology is steeped in ethical considerations and a deceptively simple aesthetic decision can easily have moral consequences (Lloyd 2009). Introducing a technology may create process efficiencies and in turn eliminate jobs. It can lower the cost of goods while promoting waste and consumerism. In playing an essential role in the process of economic and social development, introducing a new technology may produce undesirable and unintended social transformations, involving moral issues such as child labor, women's economic participation outside the home, and democracy. In such situations, objective knowledge that serves as the basis of 'right' and 'wrong' is impossible to come by because decisions of 'right' and 'wrong' are by nature ethical. It is no wonder then that this dimension of the critical reflection is so prominent, because new technologies multiply the number of ethical problems.

There is a second dimension to this critical reflection, which questions whether appropriate technology *per se* is the solution to immediate problems of poverty. This reflection does not suggest that technology is the solution *sine qua non*. Instead, it points to a deficiency in certain basic conditions that prevent the design of feasible, local solutions or technologies. For example, the absence of housing is perhaps not due to the lack of technology or materials to build houses but rather a land rights issue (Mukhija 2003). Similarly, *The Economist* reports that many mobile service innovations in Africa are constrained by a wide range of factors, including a lack of money, market knowledge, ineffective bureaucracies, taxation, and bad regulation (The Economist 2011).

---

[1] For example, Ghandi, who greatly influenced Schumacher, is said to have offered a reward for the design of a more efficient spinning wheel. However, using what (or whose?) criteria should one assess Ghandi's proposal compared to that of other advocates? What is 'efficient'? Is it the engineering definition of ratio of energy output to energy input or is it total use of resources or is it usability by a wide range of people with potential cognitive or physical disabilities? If it is efficient and requires fewer people to produce fabric, resulting in job losses, is this efficiency desirable?

This orthogonal issue is perhaps best illustrated by an extension of a well-known proverb: "Give a man a fish and he will eat for a day. Teach him how to fish and he will eat for a lifetime. *Provide him the capabilities to devise agricultural technologies and he will sustain nourishment for generations.*" In other words, we don't give the man a fishing pole, even one manufactured from locally sourced materials. Instead, we provide the man the capability to secure and protect his basic human right to nourishment through technologies of his own devising. In short, the orthogonal issue of design contributing to well-being is the issue of being capable to design, that is, having the effective opportunities to create or adapt and adopt technologies that contribute to well-being.

We will ground our discussion of this orthogonal issue in the concept of development as freedom, pioneered in the normative framework for the evaluation of well-being known as the capability approach. Over the past three decades, the capability approach has emerged as the most comprehensive normative framework for conceptualizing and assessing human development (Deneulin and Shahani 2009; Robeyns 2005; Clark 2005). A key feature of the capability approach is that development should be concerned with enhancing freedoms – the capability of individuals to live a life that they value – rather than enhancing pre-determined, utilitarian concepts of well-being such as economic output (Sen 1999). In other words, the capability approach is concerned with what people are able to do and to be, and not how satisfied they are. People can become entrained to the status quo and their conception of a valued life will be determined by what is on offer rather than what should be possible. In a direct reading of the capability approach, Oosterlaken proposes an expanded conception of design and technology contributing to well-being, as defined by the capability approach, through capability-sensitive design (Oosterlaken 2009). Without denying the relevance of design and technologies to well-being, we take a complementary path by exploring the necessary inputs to transform ideas into designed works and technologies that people need to achieve self-defined goals of well-being.

For the remainder of the paper, we will choose to use the word design in place of the design of technology. In particular, we will use the word design in the sense of a projection of possibilities, of the creation of a world that does not yet exist, rather than the popular definition of design as about giving form and style. Increasingly, the definition of design is taking on a processual interpretation rather than an object-based interpretation. The design canon is moving away from describing defined methodologies specific to a class of objects or a style and moving toward a set of domain-independent primary generators (Darke 1979). In other words, independent of the domain of the design problem, whether it is a car or a building, the process of design entails, at minimum, defining objectives and constraints, creating alternative solutions, evaluating prototypes, and detailed specification of its functionality and embodiment. This definition allows the capability approach to take design closer to Bonsiepe's ideal for design as "opening a space for self-determination, and that means ensuring room for a project of one's accord" (Bonsiepe 2006, p. 29). This self-determination is shaped by internal and external factors. To quote Kwame Anthony Appiah, "If we are authors of ourselves, it is state and society that provide us with the tools and the contexts of our authorship; we may shape our selves, but others shape our shaping" (2005, p. 156). If design contributes to making a good life, we need an ethical framework that takes us further in understanding the central ethical question of what it means for a life to be good and where design fits into this question. We propose that the capability approach provides us this framework.

11.2 The Capability Approach

Before we re-conceptualize design through the capability approach, a brief discussion of one of the core features of this normative framework, the capability set, is required. A capability set is the set of real opportunities and freedoms people have to perform any of the activities associated with well-being (Sen 1993). In selecting one or more capabilities from this set, people exercise their choice to enjoy different conceptions of well-being. This concept of sets is illustrated in Fig. 11.1. Each of the points in the diagram represents what capability theorists describe as a functioning, something that people can do ('doings') or a state that a person can choose to be in ('beings'). A group of these points is an n-tuple, which is a combination of 'doings' and 'beings'. The capability set describes the set of attainable functioning n-tuples a person can achieve. In the figure, we have labeled each n-tuple in relation to 'doings' associated with designing, such as coming up with ideas, making prototypes of alternative solutions, and evaluating each of them.

Together, the capability set constitutes different conceptions of what it means to have a good life and the list of capabilities associated with each of these conceptions. The capability set is meaningful only when all of the capabilities are equally available because capabilities reflect a person's real opportunities to choose between alternative conceptions of a good life. Each of these capabilities is in turn supported by government policies, public and private institutions, and social practices, among others. Their role is one of inputs into these capabilities. For some of these capabilities, the main input will be financial resources, but for others it can be social behavior, such as sharing household duties equally between men and women so that women can take up opportunities for economic activities outside the home.

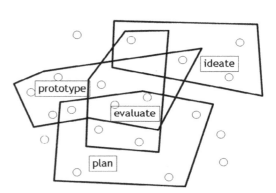

Fig. 11.1 Figurative diagram of a capability set in the field of design

Building on this notion of the capability set, we will argue that the capability approach:

1. Points to design – in the sense of envisaging and realizing a valued material world – as a central capability;
2. Suggests a multidimensional view on the capabilities to design, which is more than about augmenting the capability to design a specific object;
3. Promotes freedom in design as more than choosing between a wooden house and a brick house and hence challenges us to understand the intrinsic and instrumental freedoms of design.

We will address each of these three points in the next sections. Our aim is to illustrate why we should exert a moral concern over capability to design and consider a range of institutions and social practices to support the development of this capability.

11.3 Design as Envisaging and Realizing a Valued Material World

Specifying a list of central capabilities[2] that are associated with any reasonable conception of a life that is good is a frequently debated topic in the capability approach discourse (Robeyns 2005, p. 105). Sen subscribed to the idea of basic capabilities – capabilities that allow "a person being to do certain basic things" (1980, p. 218) – in his earlier writing (Sen 1980, 1993, p. 41). He however resists endorsing any pre-defined list of capabilities as he insists people should define their own capability sets based on their unique context (Robeyns 2005, p. 106; Clark 2005, pp. 7–8). Without taking sides in the debate whether a list of central capabilities should be prescribed, there is broad agreement that *any* list of capabilities, however derived, however contingent, must demonstrate that the capabilities in the list are central to a life worth living. In short, central capabilities are essential to any life that can flourish, no matter what a person chooses to do.

Believing that there is a normative list of capabilities for any person to have a sense of worth, it was Nussbaum (2000) who has provided the most widely discussed, and argued, list of central capabilities. For Nussbaum, central capabilities are those capabilities for which the lack of any one of the capabilities is cause for moral concern on the worth of a life. While acknowledging the contingent nature of her list, Nussbaum has strenuously defended her list of central capabilities, believing that if "people are systematically falling below the threshold in any of these core areas, this should be seen as a situation both unjust and tragic, in need of urgent attention – even if in other respects things are going well" (Nussbaum 2000, p. 71).

---

[2] The term 'capability' is used synonymously with 'capability set' in accordance with the terminology adopted by Sen.

The attraction of Nussbaum's list to our discussion is that a number of her central capabilities point to the practice of design. We emphasize again that we do not take a narrow view of design as only about form-giving, spatial layouts, or solving a problem, but rather the view that design "is concerned with how things ought to be, with devising artefacts to achieve goals" (Simon 1988, p. 69). This conception of design as being about creating the opportunities for living valued lives is consonant with several of the central capabilities on Nussbaum's list. For example, in the central capability of Senses, Imagination and Thought, Nussbaum states, "Being able to use imagination and thought in connection with experiencing and producing self-expressive works and events of one's own choice" (Nussbaum 2000, p. 78), which is linked both to the activity of ideation in design but also the outcome of designing. Nussbaum also lists the central capabilities of Control over One's (Material) Environment and Practical Reason, both of which are essential to the design activities of defining objectives and requirements and their evaluation in alternative solutions. Nussbaum explicates the capability of Practical Reason as "being able to form a conception of the good" (Ibid., p. 79). Again, this is entirely consonant with the observation that designers "bring their own intellectual program with them into each project" (Lawson 1994, p. 137) to advance the designer's vision of what the world should be like. Considered together, we interpret these three central capabilities as putatively promoting a design philosophy because they refer to envisaging and realizing a valued material world as central capabilities. Design is also instrumental in realizing other central capabilities, such as maintaining bodily integrity through shelter or other defenses of the body. Nussbaum was referring to social and political institutions when she discussed bodily integrity, but surely, adequate clothing, housing and other material forms of bodily protection play a role in bodily integrity.

In summary, the capability approach, at least as advocated by Nussbaum, points to design as a central capability. Lacking a capability to design is deprivation of a central capability, and thus design is a capability that we should exert a moral claim over. In other words, it is not the capability to design a specific object *per se* that should be the focus of our attention. Instead, it is design, which is a capability set, that is of moral import. In the next section, we turn to the multidimensional aspect of capability to address exactly this matter of being capable of 'doing' design. What is the capability set for design?

11.4 Capacity and Multidimensional Capabilities

The second theme that we explore in the concept of capability set is its multidimensional character. Before we do so, it is important to contrast between capability, which is used interchangeably with the term capacity, as adopted by the humanitarian design community, and capability as defined by the capability approach. This contrast is important, because advancing capability to design, according to the capability approach, means something qualitatively different from what is meant by advancing design capability (or building design capacity) in the humanitarian design community.

A central focus in the capacity-building approach to development is the provision of education and training to help "Strengthen capacity of primary stakeholders to implement defined activities" (Eade 1997, p. 35). In other words, capacity-building is about developing an ability or a skill. In contrast, for the capability approach, capability refers to the choice to combine capabilities, and each one of these combinations of capabilities is a realization of individual notions of well-being. A fundamental difference between capability and capacity-building is the explicit focus on the selection of one or more capabilities from a capability set in the capability approach. According to Sen, "the capability approach is ultimately concerned with the ability to achieve combinations of valued functionings" (2009, p. 233). Therefore, re-conceptualizing design through the capability approach shifts the focus to the multidimensional set of capabilities that makes design possible, which we call the capability set for design (Dong 2008).

The capability set for design is the set of capabilities associated with all of the activities associated with devising artefacts that achieve goals. We have already shown in Fig. 11.1 some n-tuples associated with design that can be chosen. Together, all of these n-tuples make up the capability to design, but people may choose alternative conceptions of design depending upon which n-tuples are chosen. The capability approach emphasizes developing capabilities that make up a functioning, rather than the functioning *per se*. For the capability approach, making design relevant to the community based on the perspectives of a 'design canon' is not the goal. In fact, a community may choose to buy all their material goods, and only apply the evaluation n-tuple in the capability set for design toward assessing how well the goods satisfy their definitions of well-being.

In saying that we are advancing capability to design, we mean that we are advancing the central capability of design, and, at the same time, we are advancing the capabilities tightly bound up with design. What then are the set of capabilities associated with design? The Design Capability Set (DCS) proposed by Dong (2008) contains one such example of a capability set for design. Dong groups capabilities in the capability set for design into six categories, or functioning n-tuples: abstraction, authority, evaluation, information, knowledge, and participation. The capabilities in these n-tuples are generated by the consideration of the activities (i.e., the 'doings') and the cognitive strategies (i.e., 'beings') commonly ascribed to the respective n-tuple. In addition to formal training, political and socioeconomic factors provide input into these capabilities.

The most relevant and valued n-tuples from the DCS proposed by Dong will vary according to context. For example, consider the windmill designed and built by William Kamkwamba to provide electricity for his village in Malawi (Kamkwamba and Mealer 2009). Kamkwamba learnt how to design the windmill capable of generating electricity by studying textbooks on physics and energy that he found in the local library. The information Kamkwamba received from the textbooks, combined with his sufficient technical knowledge in making things, and a degree of authority to realize the windmill (we could imagine how local zoning laws could have thwarted him) enabled him to design and build a technological innovation that he valued. Kamkwamba demonstrated remarkable ingenuity in adapting the information contained in the textbooks to suit local requirements and available materials. However, in many cases, people will be deprived of the capability to design for reasons beyond their immediate control. For example, should basic education such as that provided by most primary schools be deprived for economic or social reasons, an individual may be denied the opportunity to become literate. In turn, this deprivation will adversely affect their ability to make use of available information even if this information were publicly available. If the local authorities had made the permitting process to construct the windmill administratively impractical, then the deprivation is political, as we will highlight in the next example.

11.5 Multidimensional Capabilities and Political Factors

Deprivations in capability to design are also likely to stem from political factors. Let us take the specific but common situation of community engagement in the design of new public infrastructure. A range of activities comprises the design of public infrastructure, depending on the scale of the project and the ways in which it differentially impacts local communities. Activities associated with the design of public infrastructure, which could be undertaken by the community, would entail activities ranging from governance, such as approving a development proposal, to form-giving, such as participating in a design charrette. The traits of a project, such as its technical complexity or national security significance, may also alter the set of activities for which community engagement is appropriate and meaningful. The choice to 'do' these activities can only arise when we ascertain that all necessary social, political, cultural or legal machinery exists to provide the means that enable people to engage in a meaningful way. We have studied differences in capability to design for two significant public infrastructure projects of similar technical complexity and scale: the Kurnell Desalination Plant in Sydney (Australia) and the Central Freeway Replacement Project, or the Octavia Boulevard Project, in San Francisco (USA). We identified various deprivations to capabilities present in the authority, information, knowledge and participation categories in the Kurnell case which were not evident in the Octavia Boulevard case. These cases illustrate the fundamental issue that even in advanced, democratic economies, we cannot make the tacit assumption that people have the capability to design because political deprivations of capabilities indeed exist.

These examples, along with those cited earlier on land rights inhibiting housing construction in slums and the range of factors attenuating mobile phone service delivery in Africa, illustrate the difference between advancing the capability to design and building design capacity through training. In these cases, the provision of training to build design capacity is unlikely to address the underlying conditions that most inhibit the capability to design, which are likely to be political. Furthermore, in the example where training to build design capacity might have been of benefit, such as the design and construction of a windmill, it certainly was not necessary because the deprivation was in information, not knowledge. In conducting research on inequalities in capability to design, it is difficult to know *a priori* which factors have main effects in producing inequalities. Is it lack of knowledge or a land rights struggle between citizens and the government? Is it weak social institutions or strong special interests? Or, is it all of the above? A multi-dimensional capability set can help to identify sources of deprivation.

11.6 The Capability to Design and Culture

The capability set for design is only one possible multi-dimensional set. An alternative capability set for design can be derived, based on the intuitive notion that different cultures will have different understandings or perceptions on what it means to 'do' design, and thus an altogether different set of capabilities may matter when the capability set is not theorized from a Western, cognitivist perspective. This notion is essential in re-conceptualizing design through the capability approach, because different cultures have different understandings or perceptions of design if it is chosen as a functioning.

Nichols (2009) reinterpreted the categories of the Dong's (2008) DCS from the perspective of the First Australians,[3] who have already demonstrated their ability to adapt and appropriate a range of foreign technologies in valued ways that benefit their well-being, including internationally acclaimed musical innovations (for example, Yothu Yindi[4]) and automotive innovations (for example, Bush Mechanics[5]). After reviewing the relevant literature, Nichols argued that different capabilities are likely to be valued, and that the design capability set from a First Australians' perspective would focus on enhancing Indigenous knowledge systems, agency as self-determination, and local cultural vitality (Table 11.1).

We have applied these concepts on multi-dimensional capabilities through our engagement with One Laptop per Child (OLPC) Australia, a not-for-profit, non-government organization. OLPC Australia is providing children aged 4–15 living in remote areas of Australia with a connected XO laptop as part of a sustainable

---

[3] The First Australians are commonly referred to as Indigenous Australians, or Aboriginal and Torres Straight Islander peoples. It is important to note that as non-Indigenous Australians, the authors do not claim the categories identified in this chapter to be an accurate reflection of any of the Indigenous cultures in Australia. Instead, we provide them as a demonstration that we hope will trigger further discussion and debate. However, we do assume there is some common ground, and that being capable of design is important to all cultures.

[4] For general information on the band, see: http://www.yothuyindi.com/ (accessed 18 January 2011). For a detailed analysis of 'Treaty', one of Yothu Yindi's more popular songs, see Stubington and Dunbar-Hall (1994).

[5] For general information on the TV series, see: http://www.bushmechanics.com/ (accessed 18 January 2011). For a more detailed commentary of the series, see Clarsen (2002).

Table 11.1 Comparison of valued categories that support the capability to design

Dong's (2008) category	Summary of Dong's (2008) description	Summary of Nichols' (2009) First Australian reinterpretation
Abstraction	Competency in multiple levels of conceptual abstraction	Local cultural vitality with a focus on intangible artefacts such as stories and ceremony over material goods as the basis for abstraction
Authority	The power and right to enact design work	A broader sense of agency that is often expressed as collective self-determination with emphasis on maintaining harmony with others and the environment
Knowledge	Technical design knowledge	Indigenous knowledge systems in which knowledge is tied to responsibility and relationships to frame technical design knowledge

Table 11.2 Alignment of OLPC initiatives with DCS categories

Category	OLPC Australia initiative
Local cultural vitality	Localizing the operating environment of the XO laptops to include local fonts where possible (e.g. the Yolngu typeface[a] in Arnhem land)
Agency as self-determination	Introducing community controlled programs that focus on cultural learning and heritage programs to complement the predominately Western education the children receive at school
Local cultural vitality and indigenous knowledge systems	Encourage and support the development of software activities and digital content for the XO laptops that promote local language learning and other cultural knowledge

[a]http://learnline.cdu.edu.au/yolngustudies/resources_fonts.htm. Accessed 18 Jan 2011

training and support program. Many of the children who receive the XO laptops are descendants of the First Australians, providing an opportunity to investigate the applicability of Nichols' (2009) reinterpretation of the design capability set. Various deployments of the XO laptops with communities have resulted in a number of initiatives requested by the communities. In analyzing the characteristics of these initiatives, we have found that they are consonant with the type of capabilities in a capability set for design that First Australians would value, providing early confirmatory evidence of an alternative capability set for design for First Australians (Table 11.2).[6]

In cases like this where design is understood differently to that proposed by the Western design canon, the provision of training based on a Western perspective has the potential to undermine the identity of the community, and may be seen as a form of cultural imperialism or assimilation.

---

[6] At no point were the categories of the design capability set mentioned during discussions with the communities; the requests were made independently.

11.7 The Intrinsic and Instrumental Value of the Freedom to Design

The position that the capability approach takes on freedom entails expanding the definition of freedom in the direction of those positive freedoms that allow people to live a valued life. Sen attaches importance to the opportunity to pursue objectives and the process of making the choice as the type of 'real' freedoms that expand an individual's realm of achievable functionings and combinations (Sen 2009, p. 228). In other words, for Sen, freedom has both intrinsic and instrumental value. He explains this point with a comparison between human capabilities and human capital, in which he states that whilst human capital has an indirect role in influencing economic production, human capabilities have an indirect influence on social change and direct relevance to well-being and freedom. (Sen 1999, pp. 296–297) This distinction is also articulated as the difference between the means and ends of well-being and development.

In his most widely read account of the capability approach, Sen investigates five types of instrumental freedoms – political, economic, social (including education and health care), transparency (including the media) and security – and their interrelated nature:

> Freedoms are not only the primary ends of development, they are also among its principal means. In addition to acknowledging, foundationally, the evaluative importance of freedom, we also have to understand the remarkable empirical connection that links freedoms of different kinds with one another. Political freedoms (in the form of free speech and elections) help to promote economic security. Social opportunities (in the form of education and health facilities) facilitate economic participation. (Sen 1999, pp. 10–11)

We propose that design is a freedom that possesses both intrinsic and instrumental value. To have design as a freedom means *inter alia* being able to participate in the conception of the material world, having the authority to create and modify technologies, having access to the material and knowledge resources required to realize our designs, and being exposed to a variety of designed objects and the knowledge and cultural values embodied in them. We have already discussed at length about the intrinsic values of being capable of envisaging and realizing a valued material world. The instrumental value of design is it is the basis of our material culture, and therefore our quality of life by providing fundamental amenities including housing, education, recreation and community. For example, the windmill designed and built by William Kamkwamba greatly increased the quality of life for the people of his village in a range of ways, such as powering lights to allow for education opportunities, a radio and TV for information and entertainment, and recharging mobile phones for communication (Kamkwamba and Mealer 2009).

Design is the basis for Sen's other instrumental freedoms. In order to enjoy the political freedom of voting rights, suitable voting materials must be designed, including ballot papers, voter information materials, and polling booths (Lausen 2007). So that the visually impaired can live more independently and have access to economic resources, the Reserve Bank of Australia, as do many other central banks, designs currency to assist the visually impaired to recognize the different notes and coins.

The enactment of design as a freedom is perhaps best exemplified in the participatory design movement (Sanoff 2007), wherein design practitioners actively engage people in the co-creation and modification of the materials of everyday practice. Likewise in the process of co-design, in which a designer works in partnership with local residents in a process of mutual knowledge transfer and learning (Ramachandran et al. 2007). Sufficient research points to the productive outcomes from people having the freedom to design, such as squatters as developers in Mumbai (Mukhija 2003). Thus, thinking of design as a freedom, in the capability approach sense, entails consideration of the intrinsic and instrumental benefits of design in individuals' pursuit of self-defined goals and the obligation of public policies and civic administrators to promote the capability of individuals to pursue those goals through design.

11.8 Conclusions and Further Research

The capability approach has much to offer domains other than those normally considered in the humanitarian development community. This chapter has argued the three ways in which the capability approach contributes to the discussion on the ethical dimensions of design as a process. The three arguments discussed in this chapter are not meant to comprise an exhaustive list of all of the implications that re-conceptualizing design through the capability approach offers, and there are likely to be other avenues worth exploring. For example, the implication of the information base proposed by the capability approach on assessing the outcomes of design requires further inquiry. Case studies that explore how re-conceptualizing design through the capability approach translates into the nature and features of designed artefacts (such as appropriate technologies), and the manner by which these artefacts are introduced into the communities, should provide additional insights. The design process itself, especially participatory design and co-design, which have the potential to enhance the capability to design, are also possibly worthy of additional investigation.

This chapter has attempted to gently pry open the door provided by the capability approach into the multifaceted field of design, but considerable research remains to more fully explore the influence this framework on design theory and practice, especially in the humanitarian design field.

Humans have been modifying the environment to suit our needs for survival and expression since we first made tools. Our biology gives us unprecedented ability and flexibility to design our environment. Our political and social organs need to catch up.

References

Appiah, K. A. (2005). *The ethics of identity*. Princeton: Princeton University Press.
Bonsiepe, G. (2006). Design and democracy. *Design Issues, 22*(2), 27–34.
Clark, D. (2005). *The capability approach: Its development, critiques and recent advances* (Working Paper Series No. 32). Manchester: Global Poverty Research Group.

Clarsen, G. (2002). Still moving: Bush mechanics in the Central Desert. *Australian Humanities Review, 25*(March–May). Retrieved from http://www.australianhumanitiesreview.org/archive/Issue-March-2002/clarsen.html

Darke, J. (1979). The primary generator and the design process. *Design Studies, 1*(1), 36–44. doi:10.1016/0142-694x(79)90027-9.

Dong, A. (2008). The policy of design: A capabilities approach. *Design Issues, 24*(4), 76–87. doi:10.1162/desi.2008.24.4.76.

Deneulin, S., & Shahani, L. (2009). *An introduction to the human deveolpment and capability approach: Freedom and agency*. London: Earthscan.

Eade, D. (1997). *Capacity-building: An approach to people-centred development*. London: Oxfam UK and Ireland.

Kamkwamba, W., & Mealer, B. (2009). *The boy who harnessed the wind: Creating currents of electricity and hope*. New York: William Morrow.

Lausen, M. (2007). *Design for democracy: Ballot and election design*. Chicago: University of Chicago Press.

Lloyd, P. (2009). Ethical imagination and design. *Design Studies, 30*(2), 154–168. doi:10.1016/j.destud.2008.12.004.

Lawson, B. (1994). *Desingn in mind*. Oxford: Reed Eductional and Profesional Publishing Ltd.

Mukhija, V. (2003). *Squatters as developers?: Slum redevelopment in Mumbai*. Aldershot: Ashgate.

Nichols, C. (2009, September 10–12). *Design is ceremony*. Paper presented at the 2009 annual conference of the Human Development and Capability Association, Lima.

Nussbaum, M. C. (2000). *Women and human development: The capabilities approach*. New York: Cambridge University Press.

Oosterlaken, I. (2009). Design for development: A capability approach. *Design Issues, 25*(4), 91–102.

Ramachandran, D., Kam, M., Chiu, J., Canny, J., & Frankel, J. F. (2007). *Social dynamics of early stage co-design in developing regions*. Paper presented at the Proceedings of the SIGCHI conference on human factors in computing systems, San Jose, California.

Robeyns, I. (2005). The capability approach: A theoretical survey. *Journal of Human Development, 6*(1), 93–117.

Sanoff, H. (2007). Special issue on participatory design. *Design Studies, 28*(3), 213–215.

Schumacher, E. F. (1973). *Small is beautiful: A study of economics as if people mattered*. London: Vintage.

Sen, A. K. (1980) *Equality of what?* The Tanner Lecture on Human Values. Delivered at Stanford University, Stanford, USA, May 22, 1979.

Sen, A. K. (1993). Capability and well-being. In M. C. Nussbaum & A. K. Sen (Eds.), *The quality of life*. Oxford: Oxford University Press.

Sen, A. K. (1999). *Development as freedom*. Oxford: Oxford University Press.

Sen, A. K. (2009). *The idea of justice*. Cambridge: The Belknap Press of Harvard University Press.

Simon, H. A. (1988). The science of design: Creating the artificial. *Design Issues, 4*(1/2), 67–82.

Stairs, D. (2007). *Why design won't save the world*. Retrieved March 3, 2011, from http://observatory.designobserver.com/entry.html?entry=5777

Stubington, J., & Dunbar-Hall, P. (1994). Yothu Yindi's 'Treaty': *ganma* in music. *Popular Music, 13*(3), 243–259. doi:10.1017/S0261143000007182.

The Economist. (2011, January 27). *Not just talk: Clever services on cheap mobile phones make a powerful combination—especially in poor countries*. Retrieved from http://www.economist.com/node/18008202

Chapter 12
Processes for Just Products: The Capability Space of Participatory Design

Alexandre Apsan Frediani and Camillo Boano

12.1 Introduction

The relationship between *participation* and *design* has been explored in a variety of ways in theory and practice in the field of development. Especially in the context of urban development, the involvement of users in the process of design of housing or infrastructure interventions has mostly been supported as means to produce more responsive outputs. Building on the views and perspectives of local residents, participatory design methodologies have often addressed issues related to the physical properties of intervention, consulting residents about the appropriate characteristics, qualities and positioning of physical improvements. Participation has also been championed because of perceived social and institutional benefits, as engagement of users in the process of design facilitates a sense of ownership. This nurtures both the users' direct maintenance of facilities as well as their indirect involvement in using social mobilisation to persuade the relevant actors (state and private) to assume responsibility for such tasks. Through physical or social deterministic perspectives, participation is justified as an effective and cost-efficient process that enables design interventions to be responsive and well maintained through time. Such instrumental approach to participation has often been criticised by its localised character, addressing only the manifestation of urban problems, while leaving the root causes of inadequate access to services in the cities of the global south unchallenged.

The critique to participation in design in urban development relates to two schools of thought that are often unhelpfully conceptualised independently from each other: planning and design. From the planning perspective, the literature builds on the social/institutional deterministic approach to participation and calls

A.A. Frediani (✉) • C. Boano
Development Planning Unit, University College London, London, UK
e-mail: a.frediani@ucl.ac.uk; c.boano@ucl.ac.uk

for deliberation to be embedded in the process of democratisation, based on ideals of active citizenship and rights. Such perspective identifies the fundamental need to reveal normative principles that can guide the *process of design*. Meanwhile, from the design perspective, the literature has often focused on power relations to examine more critically the appropriation of the *products of design* in an urban context and their role in addressing issues of justice in the city. It is argued that democratic process do not necessarily ensure equitable and just outcomes nor challenge injustices. In other words, processes that are based on certain norms of participation have generated design outputs in an urban development context that have sustained exploitative relations of power within and among communities. Therefore writers such as Soja (2010) and Schneider and Till (2009) have been elaborating concepts (i.e. spatial justice, spatial agency) that build on the physical deterministic approach to participation by recognising the role of the built environment in challenging, sustaining or reproducing exclusionary processes in cities. Furthermore, within the literature of design, both Madanipour (1996, 2010) and Cuthbert (2007) have been contributing to the debate by breaking from the physical vs. social deterministic dichotomy on urban environments, instead furthering the re-conceptualisation of design as a socio-spatial process, arguing for an approach that builds its foundations on the understanding of not only the morphological characteristics of space, but also the societal processes that shape and support it in everyday life.

Existing literature on the capability approach has also been concerned with these issues, arguing that Amartya Sen's work on freedom can contribute to debates on the relationship between participation and design. However such literature has also reproduced this unhelpful dichotomy between process of design and product of design. From the process of design perspective, participation has been articulated as a fundamental mechanism to explore and expand capabilities. Such literature contributes to the thinking of process of design by outlining the need for participatory design strategies that reveal normative principles to guide the deliberation process (Crocker 2008); unfold and address power relations (Frediani 2010) and expand the informational base of citizens before decisions are made (Dong 2008). Meanwhile, in analysing the *product output of design* the current literature focuses on how physical products impact wellbeing (Frediani 2007; Oosterlaken 2009). Therefore, from such a perspective, the Capability Approach is used as a comprehensive evaluative framework to establish the link of physical and technological interventions with the expansion or constraint of capabilities (Oosterlaken 2009). The Capability Approach is relevant in such analysis as it incorporates an examination of the physical attributes of the product output of design *as well as* processes affecting its use and appropriation by individuals and groups (i.e. values, social norms, individual abilities, and structural processes).

This chapter aims to contribute to these debates by proposing a conceptual framework of Capability Space, which presupposes that the process and products of design collide with notions of justice, access and freedoms central to capability discourses. Grounding the argument in this body of literature, this paper articulates the emphasis on processes for just products. The result calls for a design that, borrowing from Madanipour (2010: 7) is "a dynamic multiplicity, in which city making is envisaged and organised as an inclusive and responsive process". Such a reorientation of discourses, centered on the notion of capability space, aims to define an interpretative perspective on the contemporary challenges of *design processes for just products*, as well as to enrich the practice of development practitioners dealing with spatial manifestations of injustice, complex urban challenges and spatial transformations at various scales. Hopefully such a reorientation serves to critically re-position space – its physical manifestations and processes of production – on the agenda of development interventions that seem to have forgotten the architectural and design values of their process and outcomes (at different scales), resulting in a de-spatialisation of development planning (Boano et al. 2011b).

With such problematic in mind, this paper aims to unveil the latent potential of the Capability Approach for application in a manner that addresses the full scope and sequence of design: process and product. The paper argues that, through the insertion of the notion of the production of space (Lefebvre 1991) into the Capability Approach discourse, design can be understood in a more comprehensive and integral manner within the notions of freedom and functioning. Therefore, it aims at moving from an approach that addresses design as an *instrument* for the expansion of capabilities, to one that understands design through the *lens* of capability approach.

The chapter firstly situates the current literature and practice of participatory design according to the discussions initiated in this introductory section on the dichotomy between physical vs. social/institutional deterministic approach to participation. Then, in Sect. 12.3, the chapter analyses the concept of capability to design (Dong 2008), identifying the space for contribution within the capability literature on which his article aims to build. In Sect. 12.4 of the chapter, the framework of capability space for participatory design is introduced. Then, procedural dimensions are identified in Sect. 12.5 of the chapter, aiming at examining in more detail the roles of practitioners and citizens along with their perspectives on design.

12.2 Situating Participatory Design

While the concept and applications of participation have been widely investigated in the field of political science and planning, there is a surprising lack of literature investigating the conceptual underpinnings of participatory design and its implications in terms of practice. The role of design in the context of human settlements in the global south has been interestingly explored historically by d'Auria et al. (2010), who positions participation within trends and changes on notions of human settlements from the disciplines of architecture, urbanism and planning. Similarly, Nabeel Hamdi (2004, 2010) has also been questioning the role of the practitioner in the process of participatory design by focusing on achieving change through the facilitation of community-led initiatives.

The prime reference explicitly addressing the notion of participatory design has been Henry Sanoff (2000, 2005, 2007), who in his publications has been expanding on a definition of participatory design as "an attitude about a force for change in the creation and management of environments for people" (2007: 59). Such a notion is embedded in the premises that "environment works better if citizens are active and involved in its creation and management instead of being treated as passive consumers" (2007: 59). Sanoff's works have been extremely useful in consolidating the field of participatory design by producing a series of methodologies that can be replicated and applied in the field. More recently, in his article on the 'Multiple Views of Participatory Design', Sanoff (2007) makes a crucial link between participatory design and the broader idea of deliberative democracy. However, his work often falls short in elaborating on crucial underlying concepts that are raised in his writings and that are fundamental in defining the purposes and application of participatory design (i.e. consensus, deliberative democracy, sense of community and collective intelligence). Many such issues are dealt with in a programmatic manner, rather than addressing the conceptual underpinnings and implications for practice.

Similarly, Fuad-Luke (2009) in his latest book *Design Activism* presents an interesting collection of historical and methodological arguments on the evolution of the notion of design, aiming to relate it to activism and change. It covers the activist territory, very broadly providing a variety of frameworks followed by a wide range of contemporary examples focused on the notion of participatory design and co-design. For the author participatory design and co-design are linked with activism by the proposition that "participation emancipates people by making them active contributors rather than passive recipients. It is therefore a form of design humanism aimed at reducing domination" (p. 94). However, Fuad-Luke also seems to avoid crucial questions related to the connections of design culture and techniques with the variety of discourses on participation, which contributes to the strangely apolitical nature of the book. It is true that in broader civil society, activism is arguably becoming more cultural in dealing with issues of identity rather than issues of civil rights. Yet activism's core is still highly political as it challenges the status quo, power relationships, the notion of justice and struggles for change. Thus, Fuad-Luke, like others in design culture, seems to be more comfortable within the confines of design's traditional organising principles such as usability, the design "project," human needs, and "doing good by design" or "do no harm". These concepts contrast fairly sharply with more conventional activists' language that deals with rights, struggles, grievances, claims and contestations, which more properly emerged in urban development discourses.

Despite this lack of problematisation of crucial concepts shaping the thinking and practice of participatory design, the methodologies for community engagement have been applied in an urban development context through various perspectives, approaches and purposes. The analysis of such strategies goes beyond the scope of this chapter, which aims at conveying some of the crucial limitations with the current thinking and practice of participatory design. Nevertheless, Fig. 12.1 offers a simplistic catalogue of such different strategies. It illustrates the already mentioned dichotomy between physical versus social deterministic approaches to participatory design and their ability to generate transformative change.

12 Processes for Just Products: The Capability Space of Participatory Design

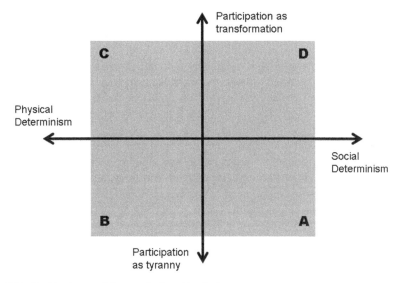

Fig. 12.1 Positioning participatory designed by Authors

Area A of the diagram represents those development initiatives that set up a series of steering committees that merely support the operational and logistical demands of projects, composed of elected leaders with no power to articulate demands or influence decision making. Normally such structures are set up in parallel with existing representative structures, often resulting in the fragmentation of social mobilisation and weakened collective action. Area B may describe projects that hold consultation events with the proposed 'beneficiaries' of projects, where professionals display often high tech and inaccessible physical plans. The motivation for such events is usually to convince people that the project displayed is the best alternative for them, rather than opening up any meaningful space for negotiation and feed-back. In both cases, the result of such a manipulative and co-opted process is frustration and increased scepticism by the general public on consultation exercises often leading to 'participation fatigue'. Examples of both types of strategies have been common in slum upgrading strategies in developing countries financed by international development agencies calling for engagement of local communities. Local governments often involve community-based organisations in slum upgrading initiatives merely with the role of supporting the implementation of pre-established plans (for a specific case study see Frediani 2007).

Areas C and D could encompass more constructive approaches to participation, where there has been a higher level of control by the general public in the process of participation. Examples of case studies in area C are participatory initiatives that use 3-D models to discuss the physical characteristics of the proposed plan (i.e. typology of houses, community facilities etc....). While participants exert a direct influence in the proposed design plan, those initiatives often provide localised solutions leaving unchallenged the structural and institutional processes underpinning the mechanisms of poverty and injustices in cities. Furthermore, as such initiatives are mostly concerned with agreeing on a final master plan acceptable by the so called 'community', little effort is given to reveal the complex power relations operating within communities as well as among the different institutional actors involved in the project. Examples of such initiatives are practices emerging from the UK-based Planning for Real methodology, which places stronger emphasis on the elaboration of collectively agreed interventions rather than revealing relations of power and causes of deprivations (for description on Planning for Real see Gibson 1994).

Precisely to respond to the limitations described above of the physical deterministic approach to participatory design, area D emerged. It includes mostly projects and cases that strengthen the capacity of grass-roots organisations in influencing policy and decision-making processes. Such an approach has often been related to methodologies of participatory enumeration and saving schemes, which support social mobilisation and enhance the bargaining power of community groups when communicating and negotiating with various actors and institutions. While having a much more explicit approach related to institutional power relations, such initiatives have paid less attention to the implications of the social production of space on the process of acquiring justice in the city. Therefore, on occasion, after long processes of institutional struggle, projects implemented on the ground potentially reproduced the same patterns of subjugation present in dominant urban development ideologies. Examples of initiatives that could be positioned in this quadrant are the practices of Shack/Slum Dwellers International (SDI), which is a transnational federation of organisations in the global south. The strategy of members of SDI involve collaboration between organisations of the urban poor and local NGOs, which provide technical assistance and attempt to build local capacities to resist evictions, survey slums and plan interventions. The SDI network has been extremely successful in creating global strategies to enhance the influence of the urban poor in decision-making. Yet when it comes to specific spatial practices, housing programmes are often still producing interventions as if they were constructed through top-down processes: high-rise buildings, sometimes in proximity of slums but also in outskirts of towns (for an example of this see the case of SPARC in Mumbai in Boano et al. (2011b, 2012)).

With such problematic in mind, this paper aims to assess the potential of the Capability Approach to be applied in a manner that addresses both elements of design: process and product. This chapter argues that by introducing the discussion on the *production of space* (Lefebvre 1991) into the Capability Approach discourse, design can be understood in a more comprehensive and integral manner through the notions of freedom and functioning. Therefore, this perspective seeks to move from an approach that addresses design as an *instrument* in the expansion of capabilities, to one that understands design through the *lens* of the capability approach. Based on the above, a participatory design framework needs to address this dichotomy between the physical and social deterministic approach embedded in a process of transformation rather than conformity or tyranny (Cooke and Kothari 2001). By building on the concept of *capability to design,* the capability approach is applied to examine the interdependent relationship between the product and process of design with freedom. For the elaboration of such analysis, the literature on the capability approach is complemented with post-structuralist discourses to deliberation and social production of space, which elaborates mechanisms to explore relations of power and identity through diversity.

12.3 Reviewing the Capability to Design

The linkages between participatory design and the capability approach have been tackled by Dong (2008), where he explores the capabilities of citizens to design, asking what is needed for the public to engage and do design. According to Dong (2008: 87):

> [T]he capability to design connects the discourse about public engagement in design to the question of who can impose order upon the designed world. If the answer to that question is the citizens who inhabit that designed world, then our attention logically turns to their capability to write and inscribe the designed world and developing their capability to express a designed world that resonates their states of mind, desires, and affects

In his article Dong seeks a normative theory on the capabilities of citizens to design which, he argues, echoes Michel Foucault's (1984) answer when asked if architecture could resolve social problems: "I think that it can and does produce positive effects when the liberating intentions of the architect coincide with the real practice of people in the exercise of their freedom". Dong (2008) examines citizens' capability to design by identifying six dimensions: information, knowledge, abstraction, evaluation, participation and authority. Dong's framework provides clear illustrations of the internal and external conversion factors influencing and shaping the realisation of such dimensions of the capability to design. Internal factors are related to, for example, the ability of designers to handle different levels of abstraction simultaneously, while external ones address the institutional arrangements that shape such abilities to design.

While the concept of 'capability to design' provides a comprehensive framework for the analysis of participatory design initiatives, there are crucial discussions and issues that would need to be addressed if the framework of capability to design is to contribute to the paradoxes between social vs. physical determinism of design and transformation vs. tyranny of participation. Although he acknowledges the element of diversity and identity shaping people's capability to design, Dong's article does not elaborate on how it actually shapes the process and values associated with design. Therefore the internal and external factors raised by Dong are mostly related to knowledge, access to information and legal frameworks while they are less about social relations and processes of domination and subordination. Due to the lack of analysis on such issues, Dong's arguments seem to suggest that if designers (professionals and community members) acquire a set of skills and inclusive legislations are developed, public engagement would be involved "effectively" in the process of design. Apart from such issues, however, it is crucial to reveal power relations and exclusionary processes operating in the context of participation that shape and influence the capability of citizens to design. Furthermore, if the perspective of capability to design takes on board issues related to power relations and exclusionary processes, it becomes crucial to discuss not only the processes that are able to reveal and challenge rather than sustain patterns of domination, but also the type of outputs generated: is participatory design concerned with consensus or discord? It is assumed in Dong's paper that if the process of deliberation is just, then agreements and consensus can be achieved through reason. However, the drive towards common views might be overshadowing difference, perpetuating the exclusion of some and consolidating the views of others, thus potentially reproducing mechanisms of oppression.

Finally, Dong's (2008) paper can be criticised for falling into the trap of social determinism, paying little attention to the physical and spatial outputs generated by the process of design. Such lack of attention for the output produced is caused by a conceptualisation of design that is merely focused on the process of deliberating on products. However, if participatory design is to move beyond the paradox of spatial versus social determinism, it needs to be embedded in a theoretical framework that is concerned not only with the process of design, but also with the characteristics, perceptions and experiences associated with the physical environment once produced. Instead of focusing on the concept of capability to design, the following part of the chapter will be building on the reflections of Dong (2008) and will examine participatory design *through* the lens of the capability approach, thus introducing the concept of capability space shaped by citizen's freedom related to the process and appropriation of products of design.

12.4 The Capability Space of Participatory Design

The first change proposed to Dong's perspective is a broadening of his focus on the capability of practitioners, to include the capability of citizens to be involved in the process of design. The six dimensions examined by Dong all relate to the skills and values that the professional should pursue in order to achieve results that would lead to human flourishing. While acknowledging the role of participation, as argued above, such perspective does not provide answers or a framework of analysis for how to engage in such a participatory process. Therefore, moving the evaluative space from practitioners to citizens, we should precisely investigate *how* citizens can be involved in design, unpacking – for example – existing design skills of citizens and their awareness of processes of domination and subordination, while also analysing local organisational capacities and assessing their ability to influence policy-making. The role of professionals would be embedded in such analysis by focusing on how they could be providing, enabling, adapting and/or sustaining (Hamdi 2010) such a citizen-centred practice of design.

Secondly, the analysis should not merely engage with the process of design, but also with its outcomes. The reason is that citizens' design freedom is shaped not merely by their choices, abilities and opportunities to engage in the process of design, but also by the degree to which the outcomes being produced are supportive of human flourishing. The capability approach has been applied to examine each of these aspects separately; to some extent it could be applied to identify some dimensions of deliberation and connected normative values like accountability, transparency, empowerment (Crocker 2008; Frediani 2010) and from there the capabilities of citizens to engage in the participatory design process can be investigated. Meanwhile, the capability approach could also be used (Oosterlaken 2009) to reveal the values associated with a product of design (i.e. the inclusiveness of a housing estate or the appropriateness of an off-grid electrical power supply system). Research should, however, be revealing how practitioners, citizens and context interact in enhancing the freedom of citizens in achieving the aspirations associated with the product of design. The challenge proposed here is thus to capture in the neologism of 'capability space of participatory design' both the process and product elements of freedom, aiming to reduce the risk of determinism dependent on their separation.

This resembles a classic debate in the field of architecture and urban design, in which authors have always challenged the physical determinism and utilitarian, functionalist perspectives embedded in a particular definition of design: the materiality of space as a societal healing machine, as a panacea for all societal problems. When only the product is taken into consideration, architecture and design are conceived and practiced through a retreat to notions of order, beauty and cleanliness. This stance can be traced to the first principles of Vitruvius, with his simplistic, but pervasive call for coherence, through to Le Corbusier, with his cry for architecture to be rid of contingent presences. Vitruvius' triad of commodity, firmness and delight remains on the architectural rosary, says Till (2009: 120), even if the beads have been updated to reflect contemporary concerns with use/function, technology/tectonics and aesthetics/beauty. The obsession with the object – the outcome – remains. When only processes are taken into consideration, architecture and design are seen as "invisible" and become marginal in its manifestation, in a way practicing what Rado Riha (2010) critically describes "architecture without architecture". This means underestimating or not even conceiving the potential that architecture and design have to "frame" space, both literally as everyday life occurs within and inside built forms, and also discursively as they transmit meanings (Dovey 1999), memories and imaginations. Finally it ignores their potential for material production and reproduction that mediate, construct and reproduce powers. An obsession with the process would mean that the importance of the outcome is overlooked.

Therefore we would like to introduce the concept of 'capability space' within the discourse on participatory design. At the centre of the concept of the capability space is the need to reveal and examine a series of normative principles that would guide the process and production of design. These normative principles are the *functionings*, in other words, as defined by Sen, the doings and beings valued by citizens. Such values are context and location specific while also relative to different moments in time (for more on different mechanisms of identifying valued capabilities see Alkire 2007). In Sect. 12.5, this chapter investigates in more detail principles that can contribute to the reflection on such values based on writings on critical theory and concerns from design experiences in contested urban areas of the global south. In relation to these functionings, the capability space emerges as citizens' freedom to pursue such values both in terms of process and product of design. Therefore the capability space is composed by these two major components: process freedom and product freedom.

Process Freedom is related to the principles and freedoms shaping deliberation, engagement and participation. This element aims at unpacking the underlying values/principles associated to participation which should guide the elaboration of methods and techniques for deliberation, which reveals, builds on and addresses the enabling and disabling conversion factors influencing the transformation of methods into achieved participatory principles. At the crux of this discussion is the understanding of participation as an end in itself, associated to a set of norms or principles and embedded in projects of deepening democracy.

Product freedom is related to the principles and freedoms shaping the production and appropriation of design outcome in a manner that supports human flourishing. This element aims at unpacking the underlying values/principles associated to design that should guide the elaboration of methods and techniques for imagining and producing spaces. At the crux of this discussion is the understanding of the tension between the potentially conflicting roles of designers as "citizen activists" on the one hand and "professional experts" on the other.

Freedom, in terms of process or product is shaped by three major elements: choice, ability and opportunity. Table 12.1 explains each of these three components of freedom in terms of process and product of design:

For the sake of conceptual synthesis, the following figure represents what we intend as the capability space of participatory design. It is composed by the process and product freedom of citizens in pursuing a certain set of valued principles associated to the deliberation and production of design.

Our critical shift reveals that design should be understood as a process, non-linear and consequential, grounded in understanding space as neither a medium nor a list of ingredients, but an interlinking of geography, built environment, symbolism, life praxis and opportunities. This type of elaboration makes explicit and intentional reference to Lefebvre's *The Production of Space* with its ontology. What profoundly challenges the notion of design from a Lefebrvian perspective is that he implies that people create the space in which they make their lives. It is not simply inherited from nature, but rather produced and reproduced through human intentions, even if unanticipated consequences develop, and even as space constrains and influences those producing it. The discovery and appropriation of such discourses makes Capability Space for design an alert to the failure of design practitioners to exercise their own critical reflections and move away from their obsession with the object, without completely dismantling the impacts of design possibilities.

In our understanding, such reformulation not only re-addresses the rupture between product and processes but potentially facilitates the creation of new forms of collective enunciations, inventing new ways of making sense of the sensible, as Rancière and Corcoran (2010: 139) puts it. Conceiving design not as imposed form or simply an aesthetic camouflage/imperative but rather as a set of practices and protocols constitutes a set of relations with time and speculative thinking. We understand participation as a design problem rather than a theoretical one, which suggests that we frame the capability space in speculative terms offered by design

Table 12.1 Components of process and product freedom

Components of freedom	Explanation in terms of process and product of design
Choice	In terms of process freedom, it is crucial that citizens' are provided with choices in relation to different mechanisms of participation (through individual consultation exercises, elected leaders, existent representatives or forums). In terms of product freedom, the provision of alternative design products (i.e. various housing typologies) is crucial to enhance citizens' freedom in participating in design
Ability	The *ability* component of the capability space is on the one hand composed by how citizens are engaged in the process of deliberation and their relationship with outside actors (i.e. architects, engineers, policy makers etc.). On the other hand, the ability component also relates to citizens' diverse capacities to appropriate the new design interventions. For example, in the context of housing, various researches (Turner 1976; Davis 2006; Frediani 2007) show that programmes often do not benefit the most vulnerable since new houses often are not affordable nor do they respond to specific needs and aspirations. Ability relates not only to individual characteristics, but also collective capacities, such as the ability to generate collective action initiatives, such as meetings and protests, as well collectively appropriating, changing, maintaining or improving the existing built environment. It is crucial therefore to accept and reveal the multiple purposes of collaboration, as it might be related to expansion of individual wellbeing or/and realisation of collective agency not necessarily generating direct benefits for individuals
Opportunity	The *opportunity* component of the capability space relates to social, economic and political processes shaping citizens' freedom to engage in the process of deliberation and appropriate products of design. In terms of process freedom, it is crucial to investigate if citizens have the role of informants, partners or leading actors of change. Therefore, it would be necessary to investigate how far participation goes: is it only about implementing a certain project in a particular locality or does it build on networks among different civil society organisations and expand room for manoeuvre to influence or change policy-making? Therefore it is fundamental to unfold the social practices shaping decision making in formal and informal institutions. Meanwhile, in the context of housing, for example, the freedom of citizens to realise their housing aspirations will be influenced by economic pressures and opportunities operating in a certain context and time, which is shaped by structural processes that go beyond the ability of individuals and groups in a locality to influence them

action rather than in the normatively certain terms offered by the perspective of participation alone.

For Bruno Latour (2004), the emergence of norms is closely connected with the movement of things from being objects or "matters of fact" to becoming "matters of concern". Becoming "matters of concern" suggests the building of new arguments for intervention, ones that pay attention to the speculative, innovative and productive potentialities of emergent collectivities rather than assuming what such collectivities are and therefore what they desire as outcomes.

12.5 Positioning the Capability Space Within Critical Thinking

If the capability space described above is to facilitate critical reflection on reality, then the linkages between process and product freedom need to be clarified in relation to wider discussions about design and participation. At the crux of these discussions are the values and motivations for design, which would allow us to examine both the role of, and the relationship between, practitioners, citizens and their respective preconceptions of design. While the literature on capability contributes to clarifying the role of process and product freedom in the discussion about participatory design, there are still many crucial design issues that are left underexplored when thinking about 'processes for just products'. Are design practitioners facilitators, providers or enablers of change? Is participatory design supposed to be striving for consensus or dissensus? Can the production of space be a means of contestating dominant exploitative ideologies? If so, how? These questions have been unpacked by a series of post-structuralist writings that examine in detail the role of power relations in the production of spaces (i.e. Lefebvre 1991; Foucault 1979; Rancière and Corcoran 2010) and processes of deliberation (hooks 1990; Gaventa 2006). What appears to interest them, especially the French philosophical circuit, and something that we feel is fundamental for our argument, is primarily the possibility of creating a methodology of interpretation of the works of art (literature first and much later architecture) capable of exposing contradictions, pluralities and disjunctions rather than proposals aimed at a unitary meaning, a totalising attempt. Such vision helps in framing Capability Space as a set of conditions to be challenged rather than a fixed given.

This section of the chapter turns now into an exploration on some of the fundamental and underlying issues and values associated to the process of participatory design based on a reflection of this literature only sketched here. Thus, four procedural dimensions are proposed to guide participatory design so that it becomes a means to generate processes for just products: Design needs to start from the position of *marginality;* to strive for *recognition* and *solidarity;* through *coalitions* based on *dissensus*. These values aim to contribute to the further exploration of the normative values identified in Fig. 12.2 that are central to the discussions on process and product freedoms.

12.5.1 Marginality: As a Starting Position

According to bell hooks, marginality is an enabling condition of resistance, and she goes on to make an important distinction between "marginality imposed by oppressive structures and the marginality one chooses as a site of resistance… as a location of radical openness and possibility" (hooks 1990:152). In hooks' work, marginality is perceived as a liberating position, one that is aware of the structural and psychic power relations shaping a particular context. For practitioners and active citizens, identifying that location of marginality means revealing the underlying processes

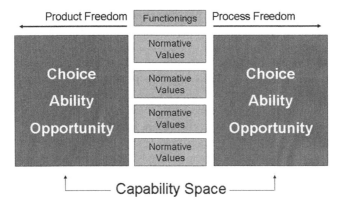

Fig. 12.2 The capability space in participatory design

that are affecting the context of participatory design, related to institutional relations, community power relations as well as inner household power struggles. Revealing marginality is about revealing conditions of subordination as well as oppression, but also spaces and opportunities for subversive thinking and practice. Therefore the subjective place of marginality is where dialogue can be established and alternative visions elaborated. If dialogue is established without contesting existing positions of marginality, than there is a great danger that denial is mirrored into conformism.

Framing design as "processes for a just product" aims to reinforce the nature of architecture and design essentially as spaces and places that challenge dominant forms of abstractions. Thus, such perspective concurs with Lefebvre's (1991) argument that there is "nothing innocent" in treating spaces as neutral containers, as such thinking actually "answers to particular strategies and tactics; it is, quite simply, the space of the dominant mode of production, and hence the space of capitalism" (1991: 360).

With such an ethical detour, Karsten Harries argues in *The Ethical Function of Architecture*, "our experience of buildings is inseparably tied to the experience we have of ourselves, of our bodies, just as our experience of our bodies is affected by the spaces we inhabit" (1997: 215). In attempting to address the ethical question, such vision remains on the level of Foucault's continual practice of freedom. Through a conscious effort to reach the other, we must be sensitive to the tastes, choices and rights of the individual user (Boano et al. 2011a, b). The commonality that Harries articulates, as a value of ethical architecture, is found at the level of a conceptual frame rather than in the final symbols or forms made legible as expressions of social structural cohesion.

Continuing to use housing as an example, following Kemeny's (1992) argument that housing is a central – if not fundamental – dimension of social structure, is the understanding of the importance of housing to society in moving away from "bricks and mortar" definitions – the physical dwelling – towards a broader social and economic dimension of built environment (Boano 2009, 2011; Boano et al. 2012). The socio-spatial relationship centred on housing can best be described in terms of "residence" which encompasses both internal dwelling and external locality factors. The physical dimension represents a conceptual move from dwelling to the aggregate level of locality. According to such a conceptualisation, housing can be approached in the broader sense of "residence" involving different levels of analysis in the interactive framework between social and physical structures. Such framework allows the understanding of built environment and housing specifically as nodes located within networks of social relations and at the centre for dynamic interplay with surrounding places (Boano et al. 2011a, b, 2012; Boano and Hunter 2012).

Therefore in the practice of participatory design, it is crucial to recognise a state of marginality: in the sense that it would mean breaking from the mainstream understanding of architecture to engage with aspirations, needs and politics related to the production of spaces; it would mean breaking from mainstream praxis of participation, questioning the logic of development practice and aid; furthermore, it would mean breaking from compartmentalised and fragmented thinking, which produces the unhelpful and problematic divisions between planning vs. design, institutional vs. spatial change, structural vs. localised practices.

12.5.2 Striving for Recognition and Solidarity

According to bell hooks (1990), the condition of marginality is understood from a position of respect, obligation and recognition of the other. Therefore it is not about separatism, but total acceptance that for the self to exist, it needs the self of others. In this sense "recognition allows a certain kind of negotiation that seems to disrupt the possibility of domination" (hooks 1994: 214). From such perspective, participatory design could be understood through a discourse of difference, diversity and interconnectivity. Design practitioners would be concerned with revealing how social identities shape and are shaped by the built environment. By unfolding social complexities of citizens involved in the design process, participants are encouraged to share diverse opinions and aspirations thus deepening the understanding of the self, norms and institutions (those being formal or informal).

Through such processes of recognition of others and affirming difference, hooks argues that solidarity is strengthened and bonding deepened. Thus, solidarity is not based on common identity or even the common history of oppression but on the transformative hope for what is yet to come. Solidarity therefore is not about merely confirming identities, but actually putting them into question and thus opening them to reconfiguration. The implications in terms of participatory design are that outputs are part of a repertoire of design possibilities, open for change and shaped through processes that are guided by normative principles, open to contestation and reconfiguration (Boano et al. 2011a, b).

In *Toward a Scientific Architecture*, argues for a replacement of the architect with a rational artisan, whose sole task would be to facilitate the user/client's wishes by building not so much form, but a repertoire of possibility. In such a move, Friedman seeks to undo an age-old architectural conundrum, that of the tension or disparity between form and use, plan and its application, architectural intellectualism with individual life. Friedman's book locates resolution within a democratisation of the planning and design process, enabling individuals to intervene within the building process with their own set of ideas, aspirations and options to determine the material conditions for their own place of residence, and stage an articulation of their own set of spatial values. The architect is thus a kind of guide, facilitating and providing knowledge of potential solutions. Such visions often used by Hamdi (2004, 2010) in its literal adoption are here to be considered not as a withdrawal from design, but a recognition of it. The architect, the design practitioner, builds a repertoire rather than the final product.

12.5.3 Coalition Through Dissensus

Participatory design is therefore based on the ideal of deepening bonds of solidarity through the recognition of social complexities and diversity. Such approach aims at revealing *dissensus,* in a Rancière sense, as a mechanism to generate coalitions present in a certain time and context which are strategic to address the causes of marginality revealed through such process. Therefore, "this connectedness does not signal a homogenous unity or a monolith totality but rather a contingent, fragile coalition building in an effort to pursue common radical libertarian and democratic goals" (West 1994: 31).

The praxis of participation based on the underlying motivation of consensus has been criticised as a mechanism to shadow conflict and perpetuate existing inequalities. Cooke (2001) uses social psychology to analyse the subtle ways in which groups make decisions to demonstrate the less visible ways that participation is used as an instrument of control and maintenance of the status quo through the production of consensus. According to Mohan (2001), "the danger from a policy point of view is that the actions based on consensus may in fact further empower the powerful vested interests that manipulated the research in the first place" (2001: 160). Finally Mosse (2001) also argues that the main limitation of participatory methods is its potential to be used as means to restrict and control the analysis of development policies: "Far from being continually challenged, prevailing preconceptions are confirmed, options narrowed, information flows into a project restricted, in a system that is increasingly controllable and closed" (Mosse 2001: 25).

Therefore, re-constructing design (urban and architectural) as a dissensus in Rancière's and Corcoran (2010) worlds seems an inevitable initial step. In similar lines, Markus Miessen's (2010) recent book on the *Nightmare of Participation* elaborates on a notion of deliberation in the process of design that contests consensus and builds on 'conflictual participation'. Conflict is not understood as a form of protest or provocation, "but rather, as micro-political practice through which the participant becomes an active agent who insists on being an actor in the force field they are facing" (2010: 93). Making references to Derrida's (1997) work on *The Politics of Friendship* and Chantal Mouffe (in Miessen 2010), Miessen argues that

Table 12.2 Dissensus design (Boano et al. 2011a)

Dissensus design	Options
Design as ANTI DESIGN (refusal)	Stemming from inappropriate design implementation, this implies not to assumingly engage in an object driven design response of 'build' and to avoid being complicit of dominant systems (economic, political, professional). This calls for abandoning craftsmanship and imaginative skills, forcing an individual or group to consider and prioritize the dynamics and processes of collective claims
Design as RESEARCH (evidence/dissensus)	Without completely abandoning creativity, imagination and craftsmanship skills, this entails making the invisible visible by employing a 'designerly' way of thinking, communicating, and reflecting that articulates and explores windows of opportunity for catalyst intervention and collectively derived design proposals within situations of uncertainty, instability and uniqueness
Design as CRITIQUE (demonstration and precedence)	Amongst debates of re-defining the urban this calls for the critical deployment of imagination and craftsmanship skills in order to question and understand complexities of contested situations. This reflective positioning offers options of speculating, mobilizing, and demonstrating the potential of informed spatial alternatives that contribute to inclusive transformation
Design as RESISTANCE	In a direct response to reducing unjust domination, there exists a condition of possibility in which design becomes a convicted emancipator using craftsmanship and imagination to promote opposition through feasible alternatives. Collectively questioning the spatial production not as objective provision, but a strategic arena for accommodating the convergence of policy, aspirations, struggles and the future

for constructive conflict to generate critical engagement there is a need to engage with 'friendly enemies' who share a symbolic space. "They [friendly-enemies] agree on the ethnic-political principles that inform the political association, but they disagree about the interpretation of these principles, a struggle between different interpretations of shared principles" (2010: 102).

However, inspired by the agonistic model developed by Mouffe as an alternative framework to cope with the stasis of consensus in political theory requires nurturing an environment, a praxis, for dissensual interaction which goes beyond a collective agreement to disagree. Reconstructing design through dissensus calls for a deeper reorientation between politics and aesthetics through what Rancière called "redistribution of the sensible" (Rancière and Corcoran 2010). In other words design dissensus provokes a disruption in the order of things, which is not simply a reordering of the relation of powers between different groups, which implies resistances, but new subjects and heterogeneous objects challenging the perception and the representation of cultural, identity and hierarchical forms. Such process of dissensus design can take different forms, from a conscious act of not intervening physically in the built environment to the production of spaces that explicitly challenges dominant ideological perspectives (Boano et al. 2011a, b; Boano and Hunter 2012). These different strategies are depicted in Table 12.2 as indication of potential 'designerly' trajectory.

12.6 Conclusion

Participatory design is about navigating through processes of transformation and tyranny. While often claimed as means to challenge exploitation perpetuated by dominant ideologies, in practice participatory design is constantly under threat to be co-opted and reproduce such processes of domination. The capability space of participatory design is proposed as a conceptual framework that hopes to guide the thinking and application of participation in the process of design. Especially in the context of urban development, conceiving design through the lens of the capability approach hopes to address the unhelpful dichotomy between design vs. planning and physical vs. social deterministic perspectives. As Blundell Jones at al. (2005: 20) remind us "[t]his is achieved through an acceptance – or let's hope a welcoming – of the political aspects of space, of the vagaries of the lives of users, of different modes of communication and representation, of an expanded definition of architectural knowledge and of the inescapable contingency of practice".

Buildings do not contain any political essence in themselves; rather they become meaningful due to the relation that their physicality has with specific social processes (Bevan 2006). Thus design should be conceived as a practice enmeshed in practices of power (Dovey 2010: 45) reinforcing the idea that it involves satisfying material needs and resolving competing social requirements through a process of active participation by the occupants and the mediation of 'professionals' (Boano 2011) who accept the responsibility of the inevitable production of identities, communities and cities – through architecture (Dovey 2010: 45).

Thus, design is simultaneously the production of physical form, the creation of social, cultural and symbolic resources and also, critically, the outcome of a facilitative process in which enablement, activism, alternatives and insurgence become central ideas. Such an approach fundamentally repositions the role of the professional. They are not, in Roy's (2006: 21) pointed phrase, the 'innocent professionals', but rather involved in a process requiring reflection upon how and what they produce paying attention to the structure of social space in what Foucault (1979) calls the 'micropractices of powers' in the everyday life. If we are to move towards more socially just and sustainable urban spaces, culturally sensitive approaches in urban transformations are a fundamental necessity. This means integrating into design (urban and architectural) the notion of casual, fluid and of course incremental production of spaces that respond to peoples' needs and aspirations, enabling sustained adaptation and fostering dissensus. Indeed, it is the task and power of design to unravel, clarify and negotiate (Boano and Hunter 2011; Boano et al. 2011a, b).

From this perspective, design, and thus the product freedoms, should be regarded as anthropocentric fields. The value of form can only be estimated against the effect these may have at a social and political level, related to the process freedom. The capability approach applied for the thinking and practice of participatory design opens up avenues to examine what design does to people's lives, unfolding its political, social, and spatial implications. The notions of functionings, capabilities and conversion factors provides a useful theoretical foundation to establish the linkages between what people value, conditions shaping their freedom to pursue such goals and design. While robust conceptually, the capability approach falls short in providing normative directions that could safeguard the project of participatory design from the threats of co-option, localism and conformity. Therefore, the capability approach is complemented with concepts emerging from critical theory and writings on radical democracy. Marginality, solidarity and coalition are applied as guiding principles in the process and product of design through participatory methods as means to address current urban challenges.

It has long been argued that the public realm of cities is in crisis, caught between privatising and commodifying tendencies and conflictive definitions. Meanwhile, design practitioners and academics search for a specific role in investigating the complexities of urbanism and in designing spaces that enable social justice and the production of alternatives towards engagement and participation. Architecture, specifically, has long lived with the tension of being seen as an art and a profession, "the most social of arts and the most aesthetic as profession" as Dovey (2010: 41) posits. As an art it carries the obligation to facilitate an inclusive, community-based imagination of transformations and changes, and as a profession to practice the public interest, with collective will and the voices of those individuals traditionally marginalised in the process.

Based on the above, design process for just product is holistic in its process and transformative in its nature, based on the critical analysis of the morphological and societal forces that shape it and the spatial political economy that governs its production. This approach inherently embeds a social, economic and political nature to the design practices which constitute real-life platforms of engagement with the contemporary urban challenges of cities in the global south. As a result, the conceptualisation and practice of design can be conceived as fostering inclusive environments through conscientiously negotiating power and authentically recognising and supporting diversity in urban space; enabling transformative processes both at the policy and at the spatial level and addressing particulars, standard enough to be applicable at different scales and flexible enough to accommodate the dynamic nature of space and social change.

References

Alkire, S. (2007). *Choosing dimensions: The capability approach and multidimensional poverty* (Chronic Poverty Research Centre Working Paper 88). OPHI, QEH, Oxford University.
Bevan, R. (2006). *The destruction of memory: Architecture at war*. Chicago: University of Chicago Press.
Blundell Jones, P., Petrescu, D., & Till, J. (Eds.). (2005). *Architecture and participation*. London: Spoon Press.
Boano, C. (2009). Housing anxiety, paradoxical spaces and multiple geographies of post tsunami housing intervention in Sri Lanka. *Disasters, 33*(4), 762–785.
Boano, C. (2011). Violent spaces: production and reproduction of security and vulnerabilities. *Journal of Architecture, 16*(1), 37–55.

Boano, C., & Hunter, W. (2011). Risks in post disaster housing: Architecture and the production of space. *ABACUS International Journal on Architecture, Conservation and Urban Studies, 5*(2). Available at http://bitmesra.ac.in/cms.aspx?this=1&mid=776&cid=560

Boano, C., Hunter, W., & Wade, A. (2011a). *BUDD design map. A visual essay*. London: Development Planning Unit. Unpublished document.

Boano, C., LaMarca, M., & Hunter, W. (2011b). The frontlines of contested urbanism mega-projects and mega-resistances in Dharavi. *Journal of Developing Societies, 27*(3&4), 295–326.

Boano, C., & Hunter, W. (2012). Architecture at risk (?): The ambivalent nature of post-disaster practice. *Architectoni.ca, 1*(1), 1–13.

Boano, C., Garcia-LaMarca, M., & Hunter, W. (2012). Mega-projects and mega-resistances in contested urbanism: Reclaiming the right to the city in Dharavi. In I. Boniburini, L. Moretto, H. Smith, & J. Le Maire (Eds.), *The right to the city. The city as common good: Between social politics and urban planning* (Cahier de la Cambre n°11) (forthcoming).

Cooke, B. (2001). The socio-psychological limits of participation? In B. Cooke & U. Kothari (Eds.), *Participation: The new tyranny?* London: Zed Books.

Cooke, B., & Kothari, U. (Eds.). (2001). *Participation: The new tyranny?* London: Zed Books.

Crocker, D. (2008). *Ethics of global development. Agency, capability, and deliberative democracy*. Cambridge: Cambridge University Press.

Cuthbert, A. (2007). Urban design: requiem for an era – review and critique of the last 50 years. *Urban Design International, 12*, 177–223.

d'Auria, V., De Meulder, B., & Shannon, K. (2010). The nebulous notion of human settlements. In B. De Meulder & K. Shannon (Eds.), *Human settlements: Formulations and (re)calibrations*. Amsterdam: SUN Architecture Publishers.

Davis, M. (2006). *Planet of slums*. London: Verso.

Derrida, J. (1997). *Politics of friendship*. London: Verso.

Dong, A. (2008). The policy of design: A capabilities approach. *Design Issues, 24*(4), 76–87.

Dovey, K. (1999). *Framing places: Mediating power in built form*. London/New York: Routledge.

Dovey, K. (2010). *Becoming places. Urbanism/architecture/identity/power*. Oxford: Routledge.

Foucault, M. (1979). *Abnormal. Lectures at the Collège de France, 1974–1975* (G. Burchell, Trans.) (p. 373). New York: Picador.

Foucault, M. (1984). Different spaces. In J. D. Faubion (Ed.), *Essential works of Foucault 1954–1984* (Aesthetic, epistemology, methodology, Vol. II). New York: New York Press.

Frediani, A. A. (2007). Amartya Sen, the World Bank, and the redress of urban poverty: A Brazilian case study. *Journal of Human Development and Capabilities, 8*(1), 133–152.

Frediani, A. A. (2010). The capability approach as a framework to the practice of development. *Development in Practice, 20*(2), 173–187.

Fuad-Luke, A. (2009). *Design activism, beautiful strangeness for a sustainable world*. London: Earthscan.

Gaventa, J. (2006). *Deepening the 'Deepening Democracy' Debate* (IDS Working Paper 264), Sussex

Gibson, T. (1994). *Showing what you mean (not just talking about it)*, PLA notes, Issue 21, pp. 41–47.

Hamdi, N. (2004). *Small change*. London: Earthscan.

Hamdi, N. (2010). *The placemaker's guide to building community*. London: Earthscan.

Harries, K. (1997). *The ethical function of architecture*. Cambridge: MIT Press.

hooks, b. (1990). Choosing the margin as a space of radical openness. In b hooks (Ed.), *Yearnings: Race, gender, and cultural politics*. Boston: South End.

hooks, b. (1994). *Outlaw culture*. London: Routledge.

Kemeny, J. (1992). *Housing and social structure. Towards a sociology of residence* (Working Paper No.12). Bristol: SAUS publication/University of Bristol.

Latour, B. (2004). Why has critique run out of steam? From matters of fact to matters of concern. *Critical Enquiry, 30*(2), 225–248.

Lefebvre, H. (1991). *The production of space* (D. Nicholson-Smith, Trans.). Oxford: Wiley-Blackwell.

Madanipour, A. (1996). *Design of urban space: An inquiry into a socio-spatial process.* Chichester: Wiley.
Madanipour, A. (Ed.). (2010). *Whose public space? International case studies in urban design and development.* London/New York: Routledge.
Miessen, M. (2010). *The nightmare of participation.* Berlin: Sternberg.
Mohan, G. (2001). Beyond participation: Strategies for deeper empowerment. In B. Cooke & U. Kothari (Eds.), *Participation: The new tyranny?* London: Zed Books.
Mosse, D. (2001). "People's knowledge", participation and patronage: Operations and representations in rural development. In B. Cooke & U. Kothari (Eds.), *Participation: The new tyranny?* London: Zed Books.
Mouffe, C. (1996). Democracy, power, and the 'political'. In S. Benhabib (Ed.), *Democracy and difference.* Princeton: University Press.
Oosterlaken, I. (2009). Design for development: A capability approach. *Design Issues, 25*(4), 91–102.
Rancière, J., & Corcoran, S. (2010). *Dissensus on politics and aesthetics.* Continuum London.
Riha, R. (2010). Architecture and new ontologies. In J. Bickert (Ed.), *Project architecture: Creative practice in the time of global capitalism.* Ljubljana: Architecture Museum of Ljubljana.
Roy, A. (2006). Praxis in the time of empire. *Planning Theory, 5*(1), 7–29.
Sanoff, H. (2000). *Community participation methods in design and planning.* New York: Wiley.
Sanoff, H. (2005). Origins of community design. *Progressive Planning, 166,* 14–17.
Sanoff, H. (2007). Multiple views on participatory design. *International Journal of Architectural Research, 2*(1), 57–69.
Schneider, T., & Till, J. (2009). Agency in architecture: Reframing critically in theory and practice, *Spring,* pp. 97–111.
Soja, E. W. (2010). *Seeking spatial justice.* Minneapolis/London: UMP.
Till, J. (2009). *Architecture depends.* Cambridge, MA: MIT.
Turner, J. F. C. (1976). *Housing by people.* London: Marion Boyars.
West, C. (1994). *Keeping faith: Philosophy and race in America.* New York: Routledge.

Chapter 13
Inappropriate Artefact, Unjust Design? Human Diversity as a Key Concern in the Capability Approach and Inclusive Design

Ilse Oosterlaken

13.1 Introduction

Sometimes the obvious is easily overlooked in our policies and practices. We all know, for example, that human beings differ from one another in countless ways; Some people live in a cold climate and others in the tropics, some people are highly educated and others are illiterate, some people are tall and others are small, some people are disabled and others are able-bodied, and so on. Yet we do not always act on it; until a few decades ago most public buildings were inaccessible to the disabled. And many websites are nowadays still hard or impossible to use by many groups, such as the blind or people with a slow and unstable internet connection (not uncommon in many developing countries). In these cases design excludes some people from reaping the benefits of the resource in question. And when design is inappropriate for some group of people, we may sometimes be looking at injustice – as the disability movement has successfully argued for buildings being wheelchair-unfriendly.

Awareness of and reflection on the importance of the fact of human diversity can, however, be found in both the literature on the capability approach and on the inclusive/universal design movement. The former is a prominent approach within political philosophy and development ethics, founded by Amartya Sen (e.g. 1999) and Martha Nussbaum (e.g. 2000). It provides a philosophical framework that has been used to think about, assess and evaluate individual well-being, as well as social arrangements and policies (Robeyns 2006). Nussbaum has formulated a list of ten categories of human capabilities to which every person has, she argues, a right. This makes the capability approach thoroughly normative, it demands – at least in Nussbaum's version of it -- (political) action aimed at bringing people to at least a

I. Oosterlaken (✉)
Philosophy Section, Delft University of Technology,
P.O. Box 5015, 2600 GA Delft, The Netherlands
e-mail: e.t.oosterlaken@tudelft.nl

certain threshold of valuable capabilities. Human capabilities and justice are thus intimately related in the capability approach. And human diversity is one of the main reasons to focus on human capabilities, instead of on the distribution of resources; the capability approach recognizes that, due to human diversity, access to a resource does not always translate in an expansion of human capabilities. An example that Sen sometimes gives is that a bicycle does not increase the things that a paralyzed person is able to do.

Human diversity is also central to the so-called 'universal design' movement, which can even be said to be all about "accounting for diversity" (Nieusma 2004, p. 14). "The discourse on universal design assumes", say Connell and Sanford (1999, p. 49), "that it is possible to design objects and spaces such that they are usable (and will be used) by a broad range of the population, including but not limited to people with disabilities." That such movements came into existence, advocating a change in design practices, indicates that designers have not always been taking human diversity *sufficiently* into account. Yet partly thanks to such social design movements, many engineers are nowadays doing so. They either design artefacts for specific, sometimes previously ignored, users (like a manually operated tricycle for disabled people in developing countries, Van Boeijen 1996), or they try to make designs that are appropriate for a wide diversity of users (such as buildings that are also accessible to wheelchairs).

The capability approach and the universal/inclusive design movement thus share important commonalities and they could benefit from each other (Oosterlaken 2009). The capability approach could learn from how the inclusive/universal design has come up with solutions for the challenge of human diversity, thus contributing to the expansion of human capabilities and the practical realization of the normative ideals of the capability approach. And the universal/inclusive design movement could benefit from a better acquaintance with the capability approach and the conceptual framework it provides. It may help designers to get a better understanding of the ultimate aims of design and may make it possible for them to make a quite natural connection between their work and wider normative debates about justice and development.

In this paper I will aim at two things, for which I will make use from insights from analytic philosophy of technology. Firstly, I would like to better explicate the intimate link between design and human capabilities and show that the apparent commonalities between the capability approach and universal/inclusive design run deeper than one might think; Using the work of Houkes and Vermaas, I will argue that the concern for human capabilities is something deeply ingrained in the nature of technical artefacts and engineering design. One difference is that engineering is concerned with expanding people's capabilities in general, irrespective of their moral value, whereas the capability approach focuses on specific individual capabilities which are held to be salient from an ethics/justice perspective.

Secondly, I will explore the grounds for moral judgments that there is sometimes injustice in an artefact by its design being inappropriate for certain groups of users. For this purpose I make use of the work of Franssen on the normativity of technical artefacts, which presents an account of statements like 'this is a good bicycle'.

Judgments like 'this is an inappropriate bicycle for a disabled person' are, however, not being discussed by him. I will thus first expand his work for the purposes of this paper. As it is exactly the fact of immense human diversity that makes inappropriateness an often occurring phenomenon, one might then say that the capability approach and the universal/inclusive design movement also have something to offer to philosophy of technology: a forceful reminder of the ubiquity and pervasiveness of the fact of human diversity, which should not be overlooked in analyses by philosophers of technology.

The set-up of this paper is as follows. I will first further explore the topic of human diversity, design and the expansion of human capabilities, using Sen's example of the bicycle (Sect. 13.2), and settle some definitional issues (Sect. 13.3). I will then discuss the work of Houkes and Vermaas on the nature of technical artefacts and engineering design and the place that human diversity has in their work (Sect. 13.4). Next, I will discuss Franssen's account of statements like 'this is a good bicycle' (Sect. 13.5) and expand it with an analysis of statements like 'this is an inappropriate bicycle for this user' (Sects. 13.6, 13.7, 13.8 and 13.9). I will then sketch how these insights from philosophy of technology provide a more in-depth understanding of the bicycle example being discussed in the literature on the capability approach (Sect. 13.10). As this approach is thoroughly normative, it is also important to be able to make the step from judgments of inappropriateness to judgments of injustice. I will argue that the capability approach can provide some normative grounding for this step (Sect. 13.11). I will end with some final reflections on further work that needs to be done on the topic of technology/design and human capabilities.

13.2 Human Diversity, Design and the Expansion of Human Capabilities

As said, human diversity forms a linking pin between the capability approach and universal/inclusive design. If design does not take facts of human diversity sufficiently into account, an artefact will not expand the capabilities of all user groups. This can be further explained using the example of a bicycle, which is also occasionally mentioned by Sen to illustrate the focus of his approach on capabilities instead of resources or utility as the space of equality:

> Take a bicycle. […] Having a bike gives a person the ability to move about in a certain way that he may not be able to do without the bike. So the transportation *characteristic* of the bike gives the person the *capability* of moving in a certain way. That capability may give the person utility or happiness if he seeks such movement or finds it pleasurable. So there is, as it were, a *sequence* from a commodity (in this case a bike), to characteristics (in this case, transportation), to capability to function (in this case, the ability to move), to utility (in this case, pleasure from moving). (Sen 1983, p. 160)

In this sequence from resource to capability to utility, it is human capabilities that are – according to Sen – the best 'space of equality'. In his book *Commodities and Capabilities* (1985, p. 9) he criticizes (welfare) economists for their tendency to view resources like bicycles "in terms of their characteristics", where "the characteristics are the various desirable properties of the commodities in question". To Sen's dissatisfaction:

> A bicycle is treated as having the characteristic of 'transportation', and this is the case whether or not the particular person happening to possess the bike is able-bodied or cripple. (Sen 1985, p. 10)

What is important according to the capability approach is rather "what the person will be able to do with those properties", or to which (human) capabilities the bicycle will contribute – none in the case of a disabled person. As people differ greatly in their personal characteristics and circumstances – Nussbaum and Sen bring it up numerous times in their work – the example is meant to illustrate a widespread phenomenon. So-called 'conversion factors' often hamper the conversion of resources into human capabilities. The specific example of the bicycle, however, has never been analyzed in more detail in the work of Sen. Philosophers of technology may want to dig deeper. They may wonder how exactly we can understand and explicate Sen's uneasiness with the way in which bicycles are being treated. I will get back to this at the end of this paper.

But if bicycles are not helpful for the disabled, how then to expand the capabilities of this group of people? Is it then a matter of providing the *right* resources, say a wheelchair instead of a bicycle? This artefact is not designed for an average person (as most bicycles are), but geared towards a specific group of 'a-typical' people, namely disabled people. This may be a solution in some cases. But sometimes, to use the vocabulary of the capability approach, other 'conversion factors' may still be such that an expansion of capabilities does not take place. To have a capability for mobility certain cultural practices and some suitable basic institutions may also be needed, for example constitutional rights that guarantee us freedom of movement. Moreover:

> Someone who is disabled and thus has an impaired capacity for movement may not benefit from a wheelchair (resource provision) unless her surroundings are adapted to allow wheelchair access (environmental change). She may also need to learn to use the wheelchair (capacity building). In this case resource provision can lead to capability expansion only if coupled with appropriate environmental and capacity interventions; the three factors are mutually interdependent. Because we are concerned with a value-based approach [i.e. the capability approach], the ultimate test of any intervention is the increase in the actual functionings that a person is able to achieve i.e. their degree of substantive freedom; interventions at the instrumental level must be cashed out in substantive terms – that is in enhanced [cap]ability to meet needs for health, knowledge, self-fulfillment, relationship to others and so on. (Johnstone 2007, p. 78)

Johnstone here mentions firstly environmental change and secondly capacity building as actions that need to supplement the provision of technological artefacts such as wheelchairs in order for them to expand capabilities. Those are certainly two important strategies. In this case the required environmental change does, of course, involve re-design, namely of buildings. It is in fact an intervention that Nussbaum (2006, p. 167) forcefully defends as a basic requirement of justice.[1] It is not hard to see,

---

[1] van den Hoven and Rooksby (2008) argue along similar lines regarding information technology.

however, that the details of design of the artefact itself, the wheelchair, may also matter. One could think, for example, of a wheelchair that is easier to operate or better able to climb curbs. Such a wheelchair would be appropriate for a wider variety of users respectively circumstances of usage. This seems to be the sort of thing that advocates of the inclusive/universal design movement would focus on in this case.

I have elsewhere introduced the phrase 'capability sensitive design' (Oosterlaken 2009) – analogue to the existing idea of 'value sensitive design' (van den Hoven 2007) – to capture the idea that it is morally desirable that engineers think about how they can contribute to the expansion of valuable human capabilities. Seen from the perspective of the capability approach, we would like engineers to design artefacts that not only *aim* at this expansion, but also *do* so in real-life situations – as far as within the sphere of influence and responsibility of engineers. The latter means at least properly taking applicable conversion factors into account, including making sure that the artefact is appropriate for the user and circumstances in question. Thus, I have argued that 'capability sensitive design' – whatever else it may be – will share characteristics with or embrace existing design movements like universal or inclusive design. From the perspective of the capability approach, such inclusive design seems to be – ceteris paribus – better than design that is not inclusive, as it will expand capabilities of more people.[2] Before discussing this in more detail, however, we need to settle some definitional issues that will prevent misunderstandings.

13.3 Distinguishing Human Capabilities and User Capacities

The terms 'capabilities' and 'capacities' are central to this article, but may mean different things to people, depending – amongst others – on their disciplinary background (like political philosophy or philosophy of technology). Hence, it seems wise to define these terms clearly. One useful place to start is Nussbaum (2000, pp. 84–85), who makes a distinction between:

(a) basic capabilities ("the innate equipment of individuals"),
(b) internal capabilities ("developed states of the person herself" – which require training, nurturing, etc.)
(c) combined capabilities ("the internal capabilities *combined with* suitable external conditions for the exercise of the function")

Nussbaum, in her various writings, has defined a list of ten central human capabilities that everybody is entitled to and that are needed to lead a flourishing human life. These capabilities belong to the third category. It is thus the combined capabilities which are the 'human capabilities' that the capability approach is ultimately concerned with as the end of development and the space of equality. We can use the

---

[2] The assumption here is that the artefact in question contributes to a valuable capability. See Sect. 13.11 for a discussion on this.

bicycle example to illustrate this. The *(combined, human) capability* relevant in this case is a person's capability to move about, to go to places where she wants to go. *External conditions* contributing to the realization of this capability may then include access to or possession of certain technologies (such as a car or bicycle). Yet certain *basic and/or internal capabilities* are also necessary for having this capability (such as control over one's legs or developed driving skills).

For matters of clarity, I will from now on consistently use the term '(human) *capabilities*' for what Nussbaum has called 'combined capabilities' (these are the capabilities that are a central concern within the capability approach) and '(user) *capacities*' for Nussbaum's 'basic and internal capabilities' (some of which will be necessary, although not always sufficient, for a technological artefact to result in the expansion of capabilities). This is, I admit, a somewhat arbitrary choice and some people may prefer – with good reasons – to use the terms 'capability' and 'capacity' the other way around. Yet this choice of words creates the conceptual clarity that I need to proceed with the main topic of this article: human diversity and the 'appropriateness' of technical artefacts.

I should also mention at this point that in developing my account of 'appropriateness', I define the term in a very specific way. In our everyday speech people may give a different meaning to the word 'appropriateness', such as cultural appropriateness.[3] Moreover, several other words exist which come also close to expressing what I mean with appropriateness, such as suitability, fitness, applicability and usefulness. Some readers may prefer to use one of these words for the state of affairs which I will label as appropriateness. My account of appropriateness is thus not simply a philosophical explication of a single word in ordinary language. What I aim to do is adding something to our understanding of technical artefacts and rendering the account of Franssen more complete in that in enables us to express normative judgments in response to facts of human diversity.

13.4 Human Diversity in Philosophy of Technology

The work of Houkes and Vermaas provides a suitable philosophical basis for developing an account of (in)appropriateness. It presents a philosophical reconstruction[4] of artefact design and usage, which they use to develop a philosophy of artefacts and a theory of technical functions. Their account

---

[3] The term will remind some people of the 'appropriate technology movement' that was especially prominent during the 1970s/1980s. However, the way 'appropriateness' has been used within this movement clearly goes way beyond the interpretation of appropriateness that I present in this article. For example, this article defines appropriateness in the context of individual, instrumental rationality in one's interaction with specific technical artefacts, not addressing broader (social) practices of deciding about and using technology. The appropriate technology movement, on the other hand, encompasses debates about technology in this much wider sense as well (see e.g. Chap. 7 of Oosterlaken, Grimshaw and Janssen in this edited volume).

[4] As it is a philosophical reconstruction, they do not aim at psychological accuracy in their description of artefact usage, nor at describing actual design practices in an empirically correct way.

challenges the metaphysical position that functions can be taken as the essences of artefacts [...] our account suggests rather that if artefacts have essences, it is that they are objects embedded in use plans.[5] (Houkes and Vermaas 2010, p. 137)

In their theory an artefact's function depends not just on the materiality of the artefact, but also on the 'use plan' that is associated with it. The concept of a use plan can be explained as follows:

> Characterizing a plan as a goal-directed series of considered actions, a use plan of object x is a series of such actions in which manipulations of x are included as contributions to realizing the given goal. (Vermaas and Houkes 2006, pp. 6–7)

This plan-based approach means that the standards of rationality apply. And:

> In a rational plan, the user believes that the selected objects are available for use – present and in working order – that the physical circumstances afford the use of the object, that auxiliary items are available for use, and that the user herself has the skills necessary for and is physically capable of using the object. (Vermaas and Houkes 2004, p. 59)

User capacities and circumstances are thus part of their action-theoretical reconstruction of the use of objects, which includes the condition "[user] u believes that his or her physical circumstances and set of skills support realizing [use plan] p" (Houkes and Vermaas 2010, p. 23).

Vermaas and Houkes characterize engineering design as being concerned with the construction of these use plans for artefacts and – when not existing yet – the artefacts required for executing them, instead of with the construction of *mere* new technical artefacts (e.g. Houkes 2008; Houkes and Vermaas 2006). Given that the standards of rationality apply to use plans, and that design is conceptualized by Houkes and Vermaas as the construction of use plans, relevant circumstances of usage and skills/capacities of the user are or should[6] – by implication – also be taken into account while designing:

> rational design requires [the designer to have] justifiable beliefs about, among other things, the users' skills, circumstances and available artefacts, just as rational use does. (Houkes and Vermaas 2010, p. 44)

And this can be quite challenging in the case of inclusive or universal design:

> the designer has to consider all prospective executings and executors of the use plan that he has constructed. Since these potential users may, for instance, have different skills and resources, aiding one group might actually decrease the chances of aiding the other group, meaning that the designer cannot consistently have skill-compatibility beliefs regarding

---

[5] They add to this that this perspective may be "profitably combined" or "supplemented with another common intuition about artefacts, namely that they are man-made objects."

[6] Depending on whether we read their account of engineering design as descriptive or normative or both. It seems that they intend it to be both descriptive and normative, since they also say that their related theory of function ascription is both (Vermaas and Houkes 2006, p. 9).

both groups. If a designer wants to make his design as *inclusive* or *universal* as possible, belief consistency makes good designing a considerable challenge. (Houkes and Vermaas 2010, p. 44)

The fact of human diversity means that most artefacts are not fully universal. If a designer does not rise to the challenge of universal design, the artefact that he designs will be inappropriate for at least one group of people. Judgments of the form 'this is an inappropriate artefact (for person p in circumstances C)' can then be made. As said, I will base my account of such judgments on the work of Franssen (2006, 2009), who in turn relates to the Houkes-Vermaas 'use plan account' of artefacts for developing his account of artefacts and normativity. I will now turn to Franssen's work.

13.5 The Goodness/Poorness of Artefacts

Franssen argues that when we say something like 'this is a good bicycle', 'this bicycle is malfunctioning' or even simply 'this is a bicycle', we are making normative judgments of a kind, namely normativity in the context of individual, instrumental rationality in our usage of technical artefacts. He explicates "how such judgments fit into the domain of the normative in general and what the grounds for their normativity are" and spells out what they mean exactly. Note that in discussing his account of such judgments, we are thus not yet speaking about moral judgments – to those we will get back in the last section of this paper.

In line with the work of Jonathan Dancy (2000), Franssen (2009, pp. 927–928) characterizes "the normative in general as being about the difference that facts about the world make to the question what to do or believe or aim for." Facts about the world, in this view, give rise to second-order normative facts about the relevance of the former facts to our deliberations. The relation between these two types of facts is, says Franssen, of a reason-giving type. On this account, evaluative statements (such as 'this artefact is good') can thus be interpreted as normative statements (i.e. reason-giving, in this case with respect to the artefact's use).[7] Let's look at a casually formulated example of what it entails to call something a good artefact:

> If we say that x is a good knife, we assume that all people who wish to cut something would agree that the features of this knife make it fit for cutting, that they would not urge you to start looking for a better knife, and so forth. (Franssen 2006, p. 50)

---

[7] Disagreements exist between philosophers on the nature of evaluative and normative statements and the relation between them. In this paper I adopt the position taken by Franssen, who himself builds on the work of Dancy. The reader is referred to the articles of Franssen (2006, 2009) for a more detailed explanation and defense of this characterization of the normative. A fuller treatment of this issue is unfortunately not possible within the scope of this article.

Facts concerning the use plan of an artefact are, according to Franssen, relevant for normative statements of the kind 'x is a good knife'. His formal analysis of the meaning of the statement 'x is a good K' reads as follows:

> 'x is a good K' expresses the normative fact that x has certain features f that make x a K and that make it the case that (1) p's wish to K and (2) the accordance with the use plan for x of (i) p's abilities, (ii) p's knowledge, and (iii) the circumstances in which p operates, jointly recommend that p uses x for K-ing. (Franssen 2009, p. 934)[8]

What this comes down to is that if conditions (1) and (2) are satisfied and if p's wish for K-ing is itself reasonable, p has a reason to use x for K-ing.[9] For any question about what to do, however, there may be multiple reasons for and against a certain action. What one should do will depend on the total balance of reasons, on the overall reason that one accordingly has. Thus:

> one cannot go further than saying that p has a reason not to use a poor x, not that p ought not to use x. However poor a K x may be, if no alternative is available p may still have a reason to use x, if only p's need for K-ing is urgent enough. (Franssen 2009, p. 934)

Note also that – in addition to the material features f that make x a K – two different kinds of first-order facts about the world are referred to in Franssen's analysis of 'x is a good K'. The *actual* capacities and circumstances of a specific user are one sort of fact about the world, the user capacities and circumstances *assumed* in the use plan for a certain artefact are another sort of fact about the world. In Franssen's analysis the goodness or poorness of K is relative to the latter only:

> Instrumental goodness may refer to properties of the user, but these are again physical properties; *they do not refer to individuating properties of the particular person who uses the artefact*. The goodness of a knife, for example, lies in the physical properties by which it enables its user to make cuts of a particular smoothness. This cannot be defined independently of the pressure that is exerted on the knife, which must fit *the average human*. A knife that is able to cut smoothly but only when pressed with a force of 100 kgf is not a good knife. (Franssen 2009, p. 937, emphasis is mine)

Note that Franssen here assumes that the use plan for a knife refers to an average human; Only a very non-average person will be able to press a knife with a force of

---

[8] Franssen decided on this way of putting after carefully considering several alternatives. One of his considerations was that p's wish to K may be unethical or otherwise unreasonable. As Franssen (2009, p. 932) put it: "it cannot be correct that Mrs. p is granted a reason to put the knife to her husband's throat merely because she wishes to do so and a knife that would do the job is available." On the other hand: "that she ought not to use the knife for cutting her husband's snoring short does not diminish in any way the goodness of the knife." In light of this difficulty, conditions (1) and (2) being met is not followed by the phrase "p has a reason to use x for K-ing", but merely by the phrase that these facts "jointly recommend that p uses x for K-ing". The phrase 'x recommends y for p' is adopted by Franssen from the work of John Broome and means that p has a reason to see to it that (if x is the case then y is the case). For a more detailed discussion of this issue, I refer the reader to Franssen.

[9] See the previous footnote.

100 kgf or 100 N. Use plans do not necessarily presuppose average users, but in practice designs are indeed often made with average people in mind. Of course plenty of artefacts – such as wheelchairs – are designed with a specific, a-typical user group in mind. But the use plan for a wheelchair still refers to the average disabled user, while within this group there may in reality be a large variety in capacities, such as strength in their arms. Franssen's analysis of 'x is a good K' applies to such cases as well.[10] We are, after all, able to distinguish good wheelchairs from poor wheelchairs. Having noticed this, we should be careful to correctly interpret Franssen (2006, p. 50) when he says that:

> If someone has a different opinion as to whether a certain [good] knife is good, it is *because* this person's abilities are atypical, she is, for example, left-handed or rheumatic or just plain clumsy, or because she has an atypical form of use in mind. (Franssen 2006, p. 50, emphasis is mine)

As most artefacts are designed with average humans in mind and assuming that most designers do a good job, Franssen's own analysis reveals that this a-typical person, saying 'x is a poor K' is in most cases uttering a normative statement that is false. Good knifes do not turn into poor knifes in the hands of a-typical people. And a high-quality, well-designed bicycle remains just that, even when owned by a disabled person. This disabled person may also be perfectly justified in drawing the conclusion that this bicycle is a good bicycle, even though she cannot use it. Yet it is understandable why a-typical people would sometimes make the mistake of calling a good artefact a poor one. I thus think that we should read Franssen's 'because' as giving an explanation for the occurrence of the mistake in judgment. We cannot take it as an endorsement of the correctness of the statement 'x is a poor K' or 'x is not a good K' in those cases.

One might object that this a-typical person is uttering an incomplete statement and that she may seem to be correct if we take it as short for 'x is a poor-K-for-me'. In everyday speech people may have the tendency to leave out the 'for me' part, especially if they feel that this is clear from the context in which they are speaking. Fair enough, but despite the usage of the same word 'poor', I will argue in the next section that this person is really making a different normative statement. To avoid confusion it would be better to use another adjective instead of poorness in those cases, and my proposal would be to use inappropriateness.

---

[10] Some artefacts (like advanced artificial limbs or even a simple set of false teeth) are even tailor-made, or designed with one very specific user in mind. In those cases as well, we are able to distinguish good design from poor design and Franssen's analysis of 'x is a good K' would work. It would simply be a matter of a use plan referring to one specific user, with his or her specific characteristics. But even for those artefacts there will be situations that call for a judgment of inappropriateness, a type of judgment that will be introduced and explicated in the next section. In fact, a good, tailor-made set of false teeth will most likely be inappropriate for all people except for the person for whom it was designed.

13.6   A First Exploration of the (In)appropriateness of Artefacts

A disabled person may have a good and possibly even a conclusive reason not to try and use a perfectly *good* bicycle, a reason that does not apply to the non-disabled person for which the bicycle was meant. If a left-handed person has to decide between using a *good* right-handed pair of scissors and an equally *good* left-handed pair of scissors, he/she has a reason to choose the latter. For a right-handed person it is the other way around. No matter how well or poorly designed, there will often be such differences in the reasons that a typical and an a-typical person have to use a certain artefact. This seems to call for a certain sort of normative judgment that reflects this difference in reason. Thus, I propose to extend the account of Franssen with another normative judgment, namely that an artefact is appropriate or inappropriate for specific users / circumstances that we find in the world.

What exactly makes 'inappropriateness' a different normative judgment than 'poorness'? Well, whether an artefact can be judged to be poor or good is, in Franssen's definition, relative to the *assumptions* about users and circumstances in the use plan. Such judgments are thus grounded in two categories of underlying first-order facts about the world, namely (a) the properties f of artefact x, in relation to (b) the abilities, knowledge and circumstances *that were assumed in the use plan* of K. A third category of first-order facts about the world does not come into play in such judgments, namely the (c) actual abilities, knowledge and circumstances of some particular person that uses or tries to use x. Remember that a disabled person could be perfectly justified in his judgment 'this is a good bicycle', even though the use plan of a bicycle assumes capacities that (s)he does not have. This judgment of instrumental goodness is thus not affected by the inability of this person to use his legs (a fact belonging to category c), whereas a judgment of inappropriateness for him/her would be grounded in this fact. Saying that 'x is an inappropriate K' – as people may casually do in everyday speech – should thus always be read as 'x is an inappropriate K for person p in circumstances C'. As judgments of (in)appropriateness and goodness/poorness are thus at least partly grounded in different first-order facts, they also correspond to different normative or second-order facts.

An example may clarify the grounds for a statement like 'this is an inappropriate car for p'. We will take the case of a good car which was – as is usually the case – designed for an average human adult (and not for Martians, children, giants or little people). Let's assume – for the sake of argument – that it has the following features (we are thus abstracting from the many other relevant features that a car has): (1) green, (2) sporty shape, (3) reliable engine, (4) high-quality brakes, (5) a distance y between the driver's seat and the brake. Which of these facts would ground a judgment that this car is a good car? Not features 1 and 2; they may be important for aesthetic reasons, but they are not relevant for instrumental goodness.[11] The judgment that a car is a good car would obviously be based on features 3 and 4, but also on 5.

---

[11] Although one might be able to come up with an exceptional, non-standard use for a car in which they would become relevant.

This last feature is relevant because the assumed user in the use plan – a normal adult – is of a certain height. Feature 5 means that a driver needs a minimum leg length to be able to push the brake while still being seated safely and with an ability to keep an eye on the road. If the distance from the seat to the break would be so big that only a giant could reach the breaks, it would not be a good car. We would not consider it to be a good car either if the distance was too small. A good car is appropriate for the intended user.

Which features would ground the judgment that the car is inappropriate for an 8-year-old child? Well, obviously feature 5 in combination with the fact that a child of this age is much smaller than the average user for which the car was designed. The child will thus have a lot of trouble hitting the brake. Whether the engine is reliable or not makes no difference for this judgment – feature 5 already gives the child *a* reason not to use the car.[12] However, as the child is not the user assumed by the use plan, this gives no ground for a judgment that the car is poor. Of course, which of the many features of an artefact ground a judgment of inappropriateness depends on the specific judgment that is made[13]; For which user, for which circumstance is the artefact said to be inappropriate? For example, before mentioned car may also be inappropriate for racing a certain rally track, but this is another judgment that would be grounded in another subset of the car's features (features not mentioned in this simplified example). The car example contains two quite straightforward and unproblematic cases of judgments of (in) appropriateness:

1. x is meant for (people like) p and p has no problem to use x for K-ing: x is appropriate for p (as before mentioned good car is for a normal adult user)
2. x is not meant for (people like) p and p has indeed problems using x for K-ing: x is inappropriate for p (as before mentioned good car is for a child or a little person)

Based on this example, one might propose that a judgment of inappropriateness for p in C can be made based merely on a comparison of (b) assumptions made in the use plan about users/circumstances and (c) the characteristics of some actual users/circumstances. If they match, we judge that the artefact is appropriate. And if not, we judge that the artefact is inappropriate. As inappropriateness then does not refer to (a) x and its properties f, this proposal may on second thought seem unsatisfactory. After all, the examples at the beginning of this section showed that judgments of (in)appropriateness have a reason-giving force. Where would this come from, if not from the object being a K with certain properties f? It is these properties that give an artefact its instrumental value. One might object to this that x's properties

---

[12] This analysis clarifies why Franssen (2006, p. 47) is correct in his casual remark, referring to a *good* car, that a 12 year old girl "definitely has a reason *not* to use the car to drive to school"; this good car is inappropriate for her.

[13] In many cases only a subset of all properties f relevant for the goodness/poorness of x will ground a specific judgment of inappropriateness for p. But an artefact can be inappropriate for p in more than one way. The more features f ground the judgment of inappropriateness, the more inappropriate x (ceteris paribus) is for p in C.

f feature indirectly in such a comparison, as the assumptions made in the use plan about users and their circumstances somehow get 'translated' into the properties of x. If all design was good design, it could indeed be as simple as that.

13.7 A More Detailed Analysis of the (In)appropriateness of Artefacts

The problem is that designers sometimes fail to make a proper 'translation' between the characteristics of their intended users and features f of x. Or, alternatively, they achieve more than was intended. This raises two additional cases (3 and 4) which are more controversial. Whether these are cases of inappropriateness, depends on the position that one takes (a or b) (Table 13.1).

If one takes position (a), one could indeed rely on merely a comparison of a real p with the assumptions made in the use plan in order to determine if x is appropriate or inappropriate. Choosing (b) means that for making judgments of (in)appropriateness one also has to take into account the properties f of x. As our interest is in judgments of (in)appropriateness as *normative, reason-giving* judgments, it makes more sense to choose for (b). This can be illustrated by an example of case (3), namely Franssen's case of a knife that "is able to cut smoothly but only when pressed with a force of 100 kgf", a pressure that an ordinary/average user could not exert. Yet – Franssen plausibly postulates – the designers of this knife had an ordinary/average user in mind when designing the knife. The heavy knife is thus a case of a design failure. Opting for (a) implies saying that the knife *is* appropriate for an average person. It seems obvious, however, that this average person also has a reason *not* to use the knife. This reason has something to do with the weight of the knife, but that fact could then only ground the judgment that the knife is a case poor design, as we have already made the choice – based on design intentions – that the knife is appropriate for an average user. However, it seems more simple and elegant to say – following (b) – that due to one of the design features f of the knife (namely its weight), the artefact is inappropriate for her (giving her a reason not to cut with it) *and because she is the intended user* this inappropriateness implies that we can also make a grounded judgment that it is a poor knife.

Table 13.1 Judgments of (in)appropriateness based on design intention versus design result

	(a) Design intention takes precedence ('meant for')	(b) Design result takes precedence ('de facto suitable for')
(3) x is meant for (people like) p, but p does not succeed to use x for K-ing	x is appropriate for p	x is inappropriate for p
(4) x is not meant for (people like) p, but nevertheless p has no problem to use x for K-ing	x is inappropriate for p	x is appropriate for p

The analysis of '*x* is an appropriate *K* for *p* in *C*' should thus include a reference to the artefact being a *K* with certain features *f*, in order to capture appropriateness in the intended *normative sense*. Furthermore, it matters not only that an artefact is a *K*, but also that it is an *operational K*; If *x* is malfunctioning, it would be odd to say that the appropriateness of *x* gives *p* a reason to use it for *K*-ing. My proposal would be:

> '*x* is an appropriate *K* for person *p* in circumstances *C*' expresses the normative fact that (1) *x* has certain features *f* that make it a *K* and (2) the relevant features *f"* of x are compatible with the characteristics of *p* and *C* that jointly make it the case that if (i) *x* is operational and (ii) *p* has a (reasonable[14]) wish to *K*, then *p* has a reason to use *x* for *K*-ing.

We could say that the explication of '*x* is a good *K*' focuses on the properties of *x*, while the explication of '*x* is an appropriate *K* for *p* in *C*' concentrates on the characteristics and/or circumstances of a specific *p*.

When reflecting on both Franssen's account of goodness and my account of 'appropriateness for *p* in *C*' together, it becomes clear that appropriateness for the intended users is a necessary, although not sufficient condition for good design. 'X is a good K' if and only if '*x* is an appropriate K for the intended user.' Whatever else may make a bicycle a good bicycle and a pair of scissors a good pair of scissors, it cannot be good unless it is at least appropriate *for the exact users and circumstances assumed in the artefact's use plan*. Franssen never gives a specification in engineering terms or otherwise what makes that an x is a good K. Such criteria will, he says, differ from artefact type to artefact type; In the case of bicycles, for example, the chain should not come of easily and in the case of scissors the blades should be sharp. But this one general criterion of good design – that the artefact should be appropriate for the specific users and circumstances assumed in the use plan – is valid for every artefact.

13.8 Goodness/Poorness, (In)appropriateness and the Balance of Reasons

What the proposed interpretation of '*x* is an appropriate *K* for *p* in circumstances *C*' allows for, is that this *x* may still be a poor *x* in other respects. In our simplified car example: the distance between the driver's seat and the brake may be compatible with *p*'s length (so in that respect the car is appropriate for *p*), but the car has an unreliable engine and lousy brakes (so the car is poor). Abstracting from this example, there are four basic possibilities:

1. *x* is a good *K* and appropriate for *p* in *C*
2. *x* is a good *K*, but inappropriate for *p* in *C*
3. *x* is a poor *K*, but appropriate for *p* in *C*
4. *x* is a poor *K* and inappropriate for *p* in *C*

---

[14] See footnote 6 on why, according to Franssen, this addition is necessary.

If p has a reasonable wish to K, in case 1 and 4 all the features f of x would be pressing on respectively the 'pro' and the 'contra' side of p's balance of reasons.[15] In case 2 and 3 there would be features pressing on each of the sides. Obviously p has on balance most reason to use artefact 1 and least reason to use artefact 4. Nothing can be said a-priory about the ranking of artefacts 2 and 3. Only if we know specific details concerning x and p/C, will we be able to make a choice. Take a pair of scissors, with three features: (a) the sharpness of its blades, (b) the tightness of the connection between both blades and (c) the shapes of its handles. It is feature (c) that makes the pair of scissors left-handed or right-handed, so either inappropriate or appropriate for a right-handed person. However, all three features are relevant to determine if a pair of scissors (whether left-handed or right-handed) is a good or a poor one. Now assume that left-handed p must choose between a good right-handed pair of scissors (case 2) and a poor left-handed pair of scissors (case 3). In that case, it seems to me, p has on balance reason to choose the former (case 2). However, this is a judgment based on personal experiences as a left-handed person with scissors, so on empirical facts. I have learned that right-handed scissors are only slightly inappropriate for me as a left-handed person and that features (a) and (b) are much more important for reaching my goal of cutting a piece of paper than (c).

In many other cases, however, it will be hard to tell if two or three should be preferred. Take a child that has not yet fully mastered the art of bicycle riding. But he wants to go for a bicycle ride anyway. Let's assume that bicycles differ in only two respects, so that there are four possible combinations coinciding with four bicycles available for the child:

1. A bicycle with a decent chain and training wheels
2. A bicycle with a decent chain without training wheels
3. A bicycle with a chain that comes of easily and with training wheels
4. A bicycle with a chain that comes of easily and without training wheels

Again, it is obvious that the child has the strongest reasons for taking bicycle 1 for a ride and the weakest reason – if any at all – for using bicycle 4. Of the two alternatives that rank in the middle, it is hard to tell which one should be preferred without learning specific details. It all depends. How easily does the chain of bicycle 3 come of? How poor are the cycling skills of the child really?

As mentioned at the end of Sect. 13.4, one could say that in a sense 'appropriate for' is really the same as 'good for'. So why propose a different term instead of 'good for'? The reason is that by calling it 'appropriate for', it becomes more

---

[15] Note that this is a simplified situation. Even artefacts that we call overall good may have some feature which we judge to be poor and the other way around. Thus, in reality even for artefacts that are good and appropriate for p in C (case 1) some feature f may be on the 'contra' side of p's balance of reasons. There exist also artefacts that are neither good nor poor., but something exactly in between. For these artefacts, about half of the features will press on the 'pro' and half on the 'contra' side of the balance of reasons.

transparent that it is possible for some *x* to be an appropriate *K* for person *p* in circumstances *C*, while it is actually not a good *K* (as several examples have illustrated). If we would choose to say instead that '*x* is a good *K* for person *p* in circumstances *C*', it would become more difficult to put aside – as we should – the (wrong) intuition that this implies '*x* is a good *K* – period!'

13.9 Extreme Inappropriateness

Table 13.2 visualizes the "hierarchy of normative facts" that Franssen (2009, p. 938) put together and indicates (last column) how I propose to expand it. In this section I will discuss the branch 'useless for K-ing' and at the same time deal with one of the possible objections to my account of (in)appropriateness for p in C. This objection against my analysis of '*x* is an (in)appropriate *K* for *p* in *C*' is that we may frown upon the thought that a disabled person missing both legs has *a* reason to use a good bicycle, no matter how insignificant in the balance of reasons compared with the strong reason not to use it because of its inappropriateness for him. Surely he has *no reason at all* to use it? He simply cannot do so. We can draw an analogy here with Franssen's analysis of '*x* is a malfunctioning K'. About malfunctioning artefacts he says that it would seem that:

> [...] whereas one may still have a reason to use a poor K for K-ing [...] one cannot have a reason to use a malfunctioning K for K-ing, since a malfunctioning K will not enable you to K in the slightest. So we would have to say that, in the case of a malfunctioning K, p has a *compelling* or *conclusive* reason not to use x for K-ing, in other words, that p *ought not* to use x for K-ing. (Franssen 2009, p. 935)

Just as you can have very good artefacts and very poor artefacts and everything in between, you can have very appropriate artefacts for *p* in *C* and very inappropriate artefacts for *p* in *C* and everything in between (for example, a left-handed pair of

Table 13.2 An extended version of Franssen's "hierarchy of normative facts"

Artefact *x*	Useful for K-ing	Working *K*	Poor *K*	Appropriate *K* for *p* in *C*
				Inappropriate *K* for *p* in *C*
			Good *K*	Appropriate K for *p* in *C*
				Inappropriate K for *p* in *C*
		Natural object that can do K-ing	Makes a good *K*	Appropriate K for *p* in *C*
				Inappropriate K for *p* in *C*
			Makes a poor *K*	Appropriate K for *p* in *C*
				Inappropriate K for *p* in *C*
	Useless for K-ing	Not a *K*		
		Malfunctioning *K*		
		Unfit *K* for *p* in *C* (because extremely inappropriate)		

scissors is only slightly inappropriate for a right-handed person). Just as in some cases one may have an overall reason to use a poor K for K-ing, one may in some cases have an overall reason to use an inappropriate K for K-ing (for example if the artefact will help you to safe somebody's life and a better or more appropriate artefact is not available[16]). And just as in the case of a malfunctioning K one has a conclusive reason not to use this K, we may argue that a paralyzed person has a conclusive reason not to use a bicycle for transportation, as the artefact is extremely inappropriate for him. He ought not to use it.

However, in the latter case a commonly used term analogous to 'malfunctioning' does not exist. One might call it 'extremely inappropriate for p in C'. To distinguish those cases in which an artefact is so extremely inappropriate for a user that conclusive reasons arise from it from those cases where an artefact is 'merely' very inappropriate for a user, we might want to introduce a new term. One candidate would be 'unusable' or 'useless'. In his 2006 article Franssen applied exactly the latter term to the sort of cases I am talking about now, cases where the circumstances of usage or the user's capacities make that a person has a compelling or conclusive reason not to use a certain artefact – which actually supports my claim that 'this is an inappropriate artefact for p in C' is a different kind of normative judgment than 'this is a poor artefact':

> there are many cases where someone has a reason not to use a particular artefact that do not involve a judgment of poor functioning or malfunctioning, although they involve normative facts of some sort: […] an electric drill is *useless* for drilling holes if there is no electric power at hand, […] (Franssen 2006, p. 48)

Yet whereas 'malfunctioning' is a term with a quite unique meaning, the term 'useless' can have a lot of different meanings. Indeed, in both articles Franssen also applies the term in another way, namely as an overarching category for objects that are either not a K or a malfunctioning K (see Table 13.2 with the 'hierarchy of normative facts' sketched earlier). In order to avoid confusion, it would thus be best to come up with a different term to apply in those cases where the circumstances of usage or the user's capacities make that a person ought not to use the artefact in question. I propose to refer to those cases as, say, cases where 'x is unfit for p in c'. This should thus be added as a third option – instead of 'extremely inappropriate for p in c' – to the category 'useless for K-ing'.[17]

---

[16] The features of the artefact that make it poor or inappropriate for you will still result in a reason against using it on the balance of reasons, but they will not put that much weight in the scales in this situation.

[17] Just like Franssen poses the question whether the transition from an extremely poor x to a malfunctioning x is discontinuous or continuous, one may wonder if the transition from extremely inappropriate to unfit is continuous or discontinuous. Proper treatment of this topic would go way beyond the scope of this article.

13.10 Sen and His Uneasiness with the Bicycle as a Transportation Device

It is now time to get back to the topics we started with, including Sen's uneasiness with the bicycle being characterized by welfare economists as a transportation device even though the artefact does not benefit the disabled. To ascribe 'transportation' as a characteristic to a bicycle is – in the language of engineers – the same as ascribing the function of transportation to it. After all, Sen is not speaking about any physical or structural properties of the bicycle, but about what it can do – providing us with transportation. Ascribing a function to an artefact is, on Franssen's analysis, making a normative statement (in the context of instrumental rationality). Sen's dissatisfaction might thus seem to suggest that he believes that this particular function ascription is incorrect. According to the Houkes and Vermaas (2010) theory of function ascription Sen would clearly be *wrong* on this. These philosophers of technology would agree with welfare economists that 'transportation' is the defining characteristic or function of a bicycle. However, Sen was obviously not intending this as a statement within the context of philosophy of technology.

A better way to understand Sen's complaint is to put it in the context of his reproaching economists for having a "commodity fetishism" (Sen 1984, 1985), in light of the fact that resources do not always lead to expanded human capabilities. It may then be asked if engineers tend to share this fetishism – it would certainly fit the stereotype image of engineers loving nuts and bolts, big machines or intricate gadgets. It would, however, not fit inclusive designers, who seem to put people central in their work. According to the Houkes-Vermaas theory as well, material artefacts are not so central to design as is often assumed. They (2010, p. 26) conceptualize designing in the technical realm as "primarily – sometimes even exclusively – constructing and communicating use plans." If the artefacts needed for the execution of the use plan do not exist yet, these need to be designed as well, but "activities that result in new material objects are a subtype of designing, called *product designing*" (Houkes and Vermaas 2010, p. 26). And as became clear in Sect. 13.5, their theory "emphasizes the 'instrumental' or 'goal-oriented' aspect of designing over its 'productive' or 'object-oriented' aspect" (Houkes 2008, p. 40). Put differently, according to this account "designers primarily aim at aiding prospective users to realize their goals" (Vermaas and Houkes 2006, p. 7). Obviously, people can only realize their goals when they have the capabilities to do so. This is what technical artefacts are supposed to do: expanding human capabilities. In other words, in the Houkes-Vermaas artefact theory the concern for *human* capabilities becomes thus something internal or inherent to the practice of engineering design itself.

If an artefact is inappropriate for some group of people, the introduction of this artefact will obviously not lead to the intended expansion of human capabilities for this group. This is what underlies Sen's dissatisfaction about how economist treat resources and what has been explicated in more detail in this paper. We should, however, also put Sen's complaint in the context of his ethical concern for justice and equality. What we should explore is how judgments of the inappropriateness of some artefact for a user group – which, as mentioned before, are normative in the context of instrumental rationality – link to *moral* judgments concerning the injustice of designed artefacts. I will discuss this in the next section.

13.11 Inappropriateness and Moral Judgments About Technology

Judgments about inappropriateness may sometimes be at the basis of judgments about injustice. The clearest example of this is perhaps that of building entrances that are *inappropriate* for handicapped users and hence exclude them from a part of public life. Inappropriateness and injustice are intimately related here. On the other hand: it seems odd, to take an extreme example, to call an atom bomb that is inappropriate for some group of users an exemplification of injustice. These examples show that my account of appropriateness cannot suffice for a *full* account of the injustice of artefacts, although I think it certainly contributes to such an account by providing part of the grounding for it. The interesting question is, then, how to further distinguish between those cases were inappropriateness is morally problematic from those in which it is not. It would be beyond the scope of this article to develop a complete and robust answer to the question which cases of inappropriateness are cases of injustice. All I can do here is make some loose suggestions.

For a case of inappropriateness to be a case of injustice the function that the artefact is supposed to fulfill should be morally salient. The capability approach could be used to bridge the gap between 'is' and 'ought' – to get from 'this artefact is inappropriate for this user group' to 'this artefact design is a case of injustice'. The key is to realize that if an artefact is sufficiently inappropriate for some user, it will not expand the human capabilities of this user. And the capability approach argues that some human capabilities have moral value and that it is a requirement of justice – at least in Nussbaum's version of the approach -- to bring each and every person up to at least a threshold level of these capabilities. We should note, however, that the capability approach would not value *each and every* capability that is expanded by a technological artefact. Nussbaum, for example, has created a list of ten capabilities that governments – according to her – ought to guarantee and promote. It is based on an ethical evaluation. Hence, she says (Nussbaum 2000, p. 83), the "capacity for cruelty, for example, does not figure on the list." This should be kept in mind when discussing, for example, the capabilities that are or are not being expanded by chemical weapons and other morally questionable technologies. Also in the case of morally far less controversial or even uncontroversial technologies, like cars and bicycles, inappropriateness for some user group or some circumstances of usage may not necessarily be a case of injustice. I think that most people would, for example, share my intuition that there is no injustice in cars not being appropriate for 8-year old children. In distinguishing morally problematic from morally unproblematic cases of inappropriateness, one of the factors that could play a role is the availability of alternatives – e.g. disabled people will find that alternatives exist for bicycles, but not for inaccessible government buildings.

Thus although all technologies are – on the account presented in this article – designed to expand *some* capabilities of people, their inappropriateness for some group of users is not always a case of injustice. In those cases where it is, one may also wonder if this also implies that one can say that the artefact *itself* is inherently morally bad. The question whether technical artefacts are morally neutral or value-laden has been much debated (e.g. Verbeek 2008; Radder 2009). Within the scope of this paper, I cannot do justice to this whole debate and fully defend my own position within it. So again, I can only give some loose suggestions; My account may seem to suggest that the injustice in those cases is indeed 'inscribed' in the artefact, as the designer has made material choices that automatically lead to the inclusion or exclusion of certain users or circumstances of usage. Technical artefacts hence do not seem to be fully value-neutral. However, we should note that inappropriateness on my account is – as has been explained – always a *relational* quality, as it only arises in relation to certain circumstances or certain users. This suggests that moral badness resulting from inappropriateness for some group of users does not *fully* reside in the artefact itself, but in the contingent combination of artefact (including use plan) and actual users/circumstances. This is, not coincidentally, in line with human capabilities – according to the literature on the capability approach – being relation in nature (Smith and Seward 2009; Oosterlaken 2011). Both human capabilities and inappropriateness only come into existence in an interplay between the person, the artefact/resource and the environment.

Although speaking about inherent moral badness of artefacts – for example, saying that a dangerous electric saw is morally bad – may be problematic, Franssen (2009, p. 948) thinks that it is "less controversial to establish the moral value of artefacts in a comparative sense only". He gives the example of two instrumentally equivalent electric saws, of which one is more dangerous than the other. According to him "we have good reason to call the more dangerous saw morally worse than the safer one." He might have a point there. To paraphrase his argument: If we have two instrumentally equivalent artefacts, one of which is appropriate for a larger group of users than the other, we may want to argue that the one which accommodates a larger group of users is the morally better one. We should note, however, that Franssen's argument only seems to work because the moral value of safety is quite uncontroversial. Hence, our variation on it would only work in cases where the (in) appropriateness is linked to a human capability that is deemed valuable, so that the exclusion becomes morally relevant. For example, when we compare a wheelchair-friendly building with another building that is the same, except that it is inaccessible by handicapped users. It seems odd again, however, to call an atom bomb that is appropriate for a larger group of users a morally better one than an atom bomb for which only some people have the required capacities to handle it. This illustrates once more that inappropriateness as discussed in this paper may provide *some* grounding for moral judgments about artefacts, although it is not sufficient on its own. We need to discuss – as capability theorists also propose – which human capabilities we have reason to value. These discussions are of great importance for engineers too.

13.12 A Final Reflection

In this paper I have used analytic philosophy of technology to show that the apparent commonalities between the capability approach and universal/inclusive design run deeper than one might think; Not only do both highlight human diversity, but moreover the concern for human capabilities is something deeply ingrained in the nature of technical artefacts and engineering design. In turn, the capability approach and the universal/inclusive design movement also have something to offer to philosophy of technology: a forceful reminder of the ubiquity and pervasiveness of the fact of human diversity, which means that statements concerning inappropriateness should be included in an account of normative statements about technical artefacts. When connecting three formerly separate fields of research – the capability approach, design and philosophy of technology – one runs the risk of remaining somewhat brief and sketchy at points. Certainly there is more that can and needs to be said about human capabilities, design and normativity. For example Franssen – and hence this article as well – focuses on normativity in the context of individual, instrumental rationality in our artefact usage. This, he acknowledges (2006, p. 926),

> does not exhaust by far the normative dimension of the practice of designing and making such artefacts — that is, the practice of engineering — nor the normative dimension of the practice of using them or deciding about them — in short, the practice of technology as a whole.

The same holds for this article. It seeks to explicate when and how individual usage of a technical artefact may or may not expand the capabilities of the user – something that will not happen if the artefact is inappropriate for the user in question. However, technology affects the capabilities of people also indirectly, through its indirect effects and its formative influence on social institutions and practices. Thus, a full analysis of the connections between technology and human capabilities would also require looking at practices of technology as a whole. This is beyond the scope of this article, but needs to be addressed in future work.

Acknowledgements This research has been made possible by a grant from NWO, the Netherlands Organization for Scientific Research. The author would like to thank several people. First and foremost Maarten Franssen for his extensive and constructive feedback at numerous occasions. His sharp and precise comments continuously challenged me to further sharpen my account of appropriateness. Secondly I would like to thank Katinka Waelbers, Sabine Roeser, Evan Selinger and three anonymous reviewers for their useful feedback on earlier drafts of this paper. Finally, Jeroen van den Hoven for his suggestion to explore if the work of Houkes and Vermaas can be used for explicating the relationship between technology and human capabilities.

References

Connell, B. R., & Sanford, J. A. (1999). Research implications of universal design. In E. Steinfeld & G. S. Danford (Eds.), *Enabling environments: Measuring the impact of environment on disability and rehabilitation*. New York: Kluwer.

Dancy, J. (2000). *Practical reality*. Oxford: Oxford University Press.

Franssen, M. (2006). The normativity of artefacts. *Studies in History and Philosophy of Science, 37*, 42–57.

Franssen, M. (2009). Artefacts and normativity. In A. Meijers (Ed.), *Handbook of the philosophy of science* (Philosophy of technology and engineering sciences, Vol. 9). Dordrecht: Elsevier.

Houkes, W. (2008). Designing is the construction of use plans. In P. E. Vermaas, P. Kroes, A. Light, & S. A. Moore (Eds.), *Philosophy and design: From engineering to architecture* (pp. 37–49). Heidelberg: Springer.

Houkes, W., & Vermaas, P. E. (2006). Planning behavior: Technical design as design of use plans. In P. P. C. C. Verbeek & A. F. L. Slob (Eds.), *User behavior and technology development: Shaping sustainable relations between consumers and technologies* (pp. 203–210). Dordrecht: Springer.

Houkes, W., & Vermaas, P. E. (2010). *Technical functions; on the use and design of artefacts*. Dordrecht: Springer.

Johnstone, J. (2007). Technology as empowerment: A capability approach to computer ethics. *Ethics and Information Technology, 2007*(9), 73–87.

Nieusma, D. (2004). Alternative design scholarship: Working towards appropriate design. *Design Issues, 20*(3), 13–24.

Nussbaum, M. C. (2000). *Women and human development; the capability approach*. New York: Cambridge University Press.

Nussbaum, M. C. (2006). *Frontiers of justice; disability, nationality, species membership*. Cambridge: The Belknap Press of Harvard University Press.

Oosterlaken, I. (2009). Design for development; a capability approach. *Design Issues, 25*(4), 91–102.

Oosterlaken, I. (2011). Inserting technology in the relational ontology of Sen's capability approach. *Journal of Human Development and Capabilities, 12*(3), 425–432.

Radder, H. (2009). Do technologies have normative properties? In A. Meijers (Ed.), *Handbook of the philosophy of science* (Philosophy of technology and engineering sciences, Vol. 9). Amsterdam: Reed Elsevier.

Robeyns, I. (2006). The capability approach in practice. *Journal of Political Philosophy, 14*(3), 351–376.

Sen, A. (1983). Poor, relatively speaking. *Oxford Economic Papers (New Series), 35*(2), 153–169.

Sen, A. (1984). *Resources, values and development*. Oxford: Blackwell.

Sen, A. (1985). *Commodities and capabilities*. Amsterdam/New York: North-Holland.

Sen, A. (1999). *Development as freedom*. New York: Anchor Books.

Smith, M. L., & Seward, C. (2009). The relational ontology of Amartya Sen's capability approach: Incorporating social and individual causes. *Journal of Human Development and Capabilities, 10*(2), 213–235.

Van Boeijen, A. G. C. (1996). *Development of Tricycle Production (DTP) in developing countries*. Paper presented at RESNA '96 Annual Conference; Exploring New Horizons… Pioneering the 21st Century, Salt Lake City.

van den Hoven, J. (2007). Ict and value sensitive design. In P. Goujon, S. Lavelle, P. Duquenoy, K. Kimppa, & V. Laurent (Eds.), *The information society: Innovations, legitimacy, ethics and democracy* (IFIP International Federation for Information Processing, Vol. 233, pp. 67–72). Boston: Springer.

van den Hoven, J., & Rooksby, E. (2008). Distributive justice and the value of information: A (broadly) Rawlsian approach. In J. van den Hoven & J. Weckert (Eds.), *Information technology and moral philosophy*. Cambridge: Cambridge University Press.

Verbeek, P. P. C. C. (2008). Morality in design: Design ethics and the morality of technological artifacts. In P. E. Vermaas, P. Kroes, A. Light, & S. A. Moore (Eds.), *Philosophy and design; from engineering to architecture*. Heidelberg: Springer.

Vermaas, P. E., & Houkes, W. (2004). Actions versus functions: A plea for an alternative metaphysics of artefacts. *The Monist, 87*(1), 52–71.

Vermaas, P. E., & Houkes, W. (2006). Technical functions: A drawbridge between the intentional and structural natures of technical artefacts. *Studies in History and Philosophy of Science, 37*, 5–18.

Author Biographies

Malik Aleem Ahmed is a PhD candidate in the Values and Technology Department (Section of Philosophy) and Infrastructure Systems & Services Department (Section of Information and Communication) at Faculty of Technology, Policy, and Management at the Delft University of Technology, the Netherlands. His areas of research interests are ICT and ethics, ICT for governance and rule of law in fragile countries. His PhD research concerns the usage of ICT for better governance in fragile countries.

Camillo Boano, PhD, is the Director of the MSc Building and Urban Design for Development at Development Planning Unit, UCL. He is an architect interested in theory and critical practices. His work and research interests are focused on urban development, design and urban transformations, shelter and housing interventions, reconstruction and recovery in conflicted areas, divided cities and natural disasters and on the linkages between society, space and built environment.

Alejandra Boni is an Associate Professor at the Department of Projects Engineering and team leader of the Group of Studies on Development, International Cooperation and Ethics at the Technical University of Valencia, where she also convenes the Master on Politics and Processes of Development. Her main areas of research are human development and the capability approach in education, planning and development aid. She is Associate Editor of the *Journal of Human Development and Capabilities*.

Mark Coeckelbergh teaches at the Philosophy Department of the University of Twente. He is the author of *Liberation and Passion* (2002), *The Metaphysics of Autonomy* (2004), *Imagination and Principles* (2007), and numerous journal articles on ethics and technology. Currently he is mainly working on philosophical issues in information technology, artificial intelligence, and robotics.

Andy Dong is an Associate Professor of Design and Computation and an Australian Research Council Future Fellow. His research is about 'design competence', the first principles of knowledge about designing. He introduced new algorithms grounded in spectral methods for dimensionality reduction, which address questions surrounding our representational mind and its capacity to invent new symbols. Alongside the cognitive and evolutionary issues, he is addressing the social and political factors that underpin the expression of design competence.

Álvaro Fernández-Baldor is Industrial Engineer (Spain), Master in Rural Development (Brazil) and PhD candidate at the Universitat Politècnica de València – UPV (Spain). He is currently working at the Center for Development Cooperation of the UPV, where he is also member of the Group of Studies on Development, International Cooperation and Ethics and lecturer in the Master on Politics and Processes of Development. The focus of his research is on technology and human development.

Alexandre Apsan Frediani, PhD, is a Lecturer in Community-led Development in the Global South at the Development Planning Unit, UCL. He is a development planner specialising in squatter settlement upgrading policies and participatory approaches to development. Areas of interest include human development, environmental justice, housing, urban development, participation and Amartya Sen's Capability Approach.

Paolo Gardoni is an Associate Professor of Civil Engineering at University of Illinois at Urbana-Champaign. His research focuses on sustainable development; reliability, risk and life cycle analysis; decision making under uncertainty; policies for natural hazard mitigation and disaster recovery; and engineering ethics. He is a member of the Technical Council on Life-Cycle Performance, Safety, Reliability and Risk of Structural Systems of the American Society of Civil Engineering (ASCE), of the International Association for Structural Safety and Reliability (IASSAR), and of the International Civil Engineering Risk and Reliability Association (CERRA).

David J. Grimshaw is Head of International Programme: New Technologies with Practical Action. He is also Visiting Professor in 'ICT for Development' at Royal Holloway (University of London) and in 'Development and Technology' at Coventry University. His latest book is an edited volume (with Shalini Kala): *The impact of information and communication technologies in Asia* (2011). He is on the Steering Group of MATTER, an "action-tank" devoted to making emerging technologies work for us all.

Jeroen van den Hoven is Professor of Moral Philosophy at Delft University of Technology and scientific director of the 3TU.Centre for Ethics and Technology. He specializes in the ethics of information technology. His many publications in this area include the edited volume *Information Technology and Moral Philosophy* (Cambridge University Press, together with John Weckert). He is project leader of the multidisciplinary research project 'Technology for Development – A Capability Approach', funded by the Netherlands Organization for Scientific Research.

Andrés Hueso is Mechanical Engineer and Master in Development Politics and Processes at the Universitat Politècnica de València (Spain), where he is currently PhD candidate and member of the Group of Studies on Development, Cooperation and Ethics. He has research experience in Bolivia on community hydro power projects, and on participatory governance in Angola. His current research interests are technology and human development, sustainability and sanitation.

Pim Janssen graduated in 2010 from the master program Philosophy of Science, Technology and Society at the University of Twente. For his master thesis *Kamuchina Kemombe: Opening the Black-box of Technology within the Capability Approach* he did fieldwork on an ICT project in Zimbabwe. He is currently designer with consultancy agency ARCADIS.

Kim Kullman MA is PhD Student at the University of Helsinki. He is currently writing an ethnography on the relations between young children and everyday traffic. His work has been published in various journals, including Space & Culture, Social & Cultural Geography and Children's Geographies.

Nick Lee PhD FRSA is Associate Professor at the University of Warwick. He is author of *Childhood and Society: Growing up in an Age of Uncertainty* (2001*), Childhood and Human Value: Development, Separation and Separability* (2005) and numerous articles on human and non-human agency, technology and bio-politics. He is currently writing *Childhood and Bio-Politics: Climate Change, Bioscience and Human Futures* (2011, Palgrave Macmillan) and coordinating an ESRC/BBSRC/MRC funded interdisciplinary research network focusing on mimetic factors in individual health behaviour.

Manu V. Mathai is a visiting Assistant Professor at Rochester Institute of Technology. His research and teaching interests include energy and environmental policy, and technology, environment and society studies, especially technological systems as mediators of power and possibilities for change in society. His dissertation researched India's economic development discourse and civilian nuclear power using these ideas. Manu has been an early student of extending the Human Development and Capability Approach to address the environmental crisis.

Colleen Murphy is an Associate Professor of Philosophy at University of Illinois at Urbana-Champaign. Her research focuses on the moral dimensions of those disruptions to political communities that arise from civil conflict, repression, war, and natural disasters. She is the author of *A Moral Theory of Political Reconciliation* (Cambridge University Press, 2010) has also published articles in journals including the *Journal of Risk Research, Law and Philosophy, Res Publica, Risk Analysis, and Science* and *Engineering Ethics*.

Crighton Nichols is a PhD Candidate at the Design Lab, University of Sydney, and Community Engagement and Localisation Manager at One Laptop per Child Australia. At the intersection of his research and practice is a desire to understand design at the cross-cultural interface, in order to promote the design of innovative uses and encourage the appropriation of foreign technologies by the First Australians.

Ilse Oosterlaken is a PhD candidate at the philosophy section of TU Delft / the 3TU.Centre for Ethics and Technology, as part of a multidisciplinary project on the capability approach, technology and design. Two of her recent articles are *Design for Development – A Capability Approach* (2009) and *Inserting Technology in the Relational Ontology of Sen's Capability Approach* (2011). She organized the 2011 conference of the Human Development and Capability Association, with 'innovation' as its theme.

Bernd Carsten Stahl is Professor of Critical Research in Technology and Director the Centre for Computing and Social Responsibility at De Montfort University, Leicester, UK. His interests cover philosophical issues arising from the intersections of business, technology, and information. This includes the ethics of computing and critical approaches to information systems.

Yingqin Zheng is Lecturer in Information Systems at the School of Management, Royal Holloway, University of London and Research Associate at London School of Economics (LSE). She holds a PhD and an MPhil from the University of Cambridge and an MSc from LSE. Her research interests encompass the implications of information and communication technology in society in relation to human development, including the application of Sen's capability approach in this area.

Author Index

A
Ahmed, M.A., 12, 153–168
Alampay, E.A., 155
Alampay, E.M., 11, 12
Alkire, S., 4, 6, 7, 19, 34, 90, 118, 141–144, 154, 164, 211
An, D., 183
Anderson, E., 184
Appiah, K.A., 191
Ara, R., 115
Arnold, M., 57
Asveld, L., 9

B
Bailur, S., 11, 13, 20, 114, 129, 130
Balachandra, P., 91, 92, 99
Ballet, J., 144
Barja, G., 11
Basalla, G., 32
Bates, F., 183
Bedford, T., 176, 184
Behari, B., 137
Beitler, D., 51
Bevan, R., 219
Bijker, W.E., 107
Birdsall, W.F., 7, 14
Blundell Jones, P., 219
Boano, C., 203–220
Boenink, M., 82
Boni, A., 135–150
Bonsiepe, G., 191
Borenstein, J., 10, 12
Bouton, D., 104
Bowker, G.W., 28

Boyle, T., 104
Brandão, F.C., 137
Brey, P., 63, 66
Bridges, D., 142
Buchanan, R., 64
Bull, M., 40, 48
Bunker, S.G., 88
Byrne, J., 88, 90, 103–105, 107

C
Cadili, S., 60
Callon, M., 52
Canny, J., 200
Casini, G., 157, 158, 160, 163, 167
Chambers, R., 118, 138
Chang, S.A., 103
Chatterjee, P., 94
Chiu, J., 200
Clague, J., 10
Clark, D.A., 49, 50, 52, 191, 193
Clarsen, G., 197
Coeckelbergh, M., 8, 10, 12, 16, 77–86
Coles, A., 66
Collingridge, D., 67
Comim, F., 6, 90, 144
Connell, B.R., 226
Cooke, B., 62, 208, 217
Cooke, E.F., 10
Cooke, R., 176, 184
Coombs, C.R., 60
Corcoran, S., 218
Cormenzana, F., 51
Cozzens, S.E., 10, 11

Crocker, D.A., 5, 6, 50, 65, 118, 142, 143, 179, 204, 211
Cunningham, L., 183
Cuthbert, A., 204

D
Dada, D., 153
Dagnino, R., 137, 138
Dahlhamer, J., 183
Daly, H.E., 88
Dancy, J., 232
Danowitz, A.K., 122
Darke, J., 191
Darlington, J.D., 183
Dash, N., 183
d'Auria, V., 205
Davis, M., 213
Decker, M., 74
Deitz, T., 105
De Meulder, B., 205
Deneulin, S., 6, 7, 62, 90, 101, 102, 141, 144, 191
De', R., 155
Der Kiureghian, A., 181
Derrida, J., 217
de-Shalit, A., 177, 178, 184
Dewey, J., 31
Dinda, S., 88
Ditlevsen, O., 181
Dixit, S., 92
Doherty, N.F., 60, 62
Dolsak, N., 105
Dong, A., 9, 18, 195, 198, 204, 205, 209, 210
Dovey, K., 211, 219, 220
Dreze, J., 95, 96
D'Sa, A., 91, 92, 99
D'Souza, M., 183
Dubois, A., 141
Dubois, J.L., 144
Dufumier, M., 135
Dunbar-Hall, P., 197
Duraiappah, A.K., 97

E
Eade, D., 195
Edwards, E., 183
Ehrhardt-Martinez, K., 104
Elder-Vass, D., 124
Esteva, G., 95
Evans, P., 144

F
Feenberg, A., 62
Fernández-Baldor, Á., 135–150
Fichtelius, E., 164, 165
Flores, P., 51
Foot, P., 31
Forsberg, A.B., 167
Foster, J.E., 14
Fothergill, A., 183
Foucault, M., 63, 209, 214, 219
Frankel, J.F., 200
Franssen, M., 232–234, 236, 240, 241, 244
Frediani, A.A., 9, 118, 203–220
Friedmann, J., 30, 216, 217
Fuad-Luke, A., 206
Fukuda-Parr, S., 141

G
Garai, A., 11
Garcia-LaMarca, M., 205, 208, 215, 216, 218, 219
Gardoni, P., 10, 173–185
Garnham, N., 7, 11, 154
Gasper, D., 143
Gatchair, S., 10, 11
Gaventa, J., 214
Genus, A., 66
Gibson, T., 208
Gichoya, D., 163
Gigler, B.-S., 8, 11, 17, 22
Gingrich, S., 88
Girard, C., 183
Gladwin, H., 183
Glover, L., 105, 107
Goldberg, N., 74
Goldemberg, J., 90, 91
Goodman, L.J., 137
Goodman, S.E., 122
Gordon, P., 183
Gregory, J., 104
Griffith, J., 157, 158, 160, 163, 167
Grimshaw, D.J., 113–132
Grunfeld, H., 11, 12
Grunwald, A., 74
Gudza, L.D., 115, 116, 121
Gupta, A., 108

H
Haberl, H., 88
Habermas, J., 88
Haimes, Y.Y., 184

Hak, S., 11, 12
Hamdi, N., 205, 210, 217
Hamel, J.-Y., 7
Handy, C., 14
Hansson, S.O., 184
Harries, K., 215
Harris, C.E., 176, 177
Härtel, H., 51, 54
Hassard, J., 41
Heeks, R., 12, 158
Hellsten, S.K., 12, 14
Hermant, E., 40
Herrera, A., 137–139
Herring, H., 88
Hirschheim, R., 60
Hollow, D., 51, 54
hooks, b., 214, 216
Houkes, W., 231, 232, 242
Hourcade, J.P., 51
Huang, Q., 181
Hudson, A., 163, 166
Hueso, A., 135–150
Hughes, T.P., 107
Hunter, W., 205, 208, 215, 216, 218, 219
Hurlebaus, S., 181

I

Ibrahim, S., 6, 143–145
Introna, L.D., 63

J

James, B., 135
James, J., 11, 46, 47, 54
Janssen, M., 168
Janssen, P., 113–132
Johansson, T.B., 90, 91
Johnsson, A.B., 160
Johnstone, J., 9, 11, 14, 20, 21, 34, 113, 154, 228

K

Kajitani, Y., 183
Kamkwamba, W., 195, 199
Kam, M., 200
Kemeny, J., 215
Keval, N., 62
Kjolberg, K.L., 74
Kleine, D., 11–13, 114, 128–130, 154, 155
Klein, H.K., 60, 65
Kosambi, D.D., 108

Koshal, M., 108
Koshal, R.K., 108
Kothari, U., 62, 208
Krausman, F., 88
Kroes, P., 20
Kuklys, W., 144
Kullman, K., 39–55
Kumar, A., 103
Kumar, K., 136
Kumar, R., 180
Kurdgelashvili, L., 103

L

Ladikas, M., 74
Laitner, J.A., 104
LaMarca, M., 205, 208, 215, 216, 218, 219
Latour, B., 8, 40, 41, 43, 44, 52, 53, 213
Lausen, M., 199
Law, J., 41
Lawson, B., 194
Leach, M., 138, 139
LeClair Paquet, B., 205, 208, 215, 216, 218, 219
Lee, A.S., 60
Lee, N.M., 39–55
Lefebvre, H., 205, 208, 214, 215
Leston-Bandeira, C., 160
Linn, C., 104
Lloyd, P., 190
Loan-Clarke, J., 60
Lobo, M., 137, 138
Lokhorst, G.-J., 35
Lormand, S., 183
Lovins, A.B., 107
Lowe, L., 115, 116, 121
Lyytinen, K., 60

M

Madanipour, A., 204, 205
Madon, S., 12, 154, 155
Madsen, H.O., 181
Maestas, E.G.M., 183
Mahieu, F., 144
Mainster, J., 183
Mansell, R., 11, 154, 155
Martinez, C., 90, 103–105, 107
Massey, D., 54
Mathai, M.V., 87–109
Mbarika, V., 154
McAulay, L., 62
Mealer, B., 195, 199

Meijers, A., 20
Meso, P., 154
Miessen, M., 217
Mika, L., 115–117, 127
Mileti, D.S., 183
Miller, S., 32
Mohan, G., 217
Moore, J.E.II, 183
Morris, C.W., 41
Morrow, B.H., 183
Mosalam, K.M., 181
Mosse, D., 53, 217
Mouffe, C., 217
Mukhija, V., 190, 200
Mumford, L., 87–89, 106–108
Munasinghe, M., 88
Murphy, C., 10, 173–185
Musa, P.F., 11, 154
Myers, M.D., 65

N
Nassef, Y., 122
Negroponte, N., 45
Ngwenyama, O.K., 60
Nichols, C., 189–200
Nieusma, D., 34, 119, 226
Nigg, J., 183
Norgaard, R.B., 88
Norgard, J.S., 88
Nunez Ferrera, I., 205, 208, 215, 216, 218, 219
Nussbaum, M.C., 4, 5, 10, 78–80, 85, 90, 118, 124, 144, 154, 155, 163, 177, 181, 193, 194, 225, 228, 229, 243

O
Okada, N., 183
Oosterlaken, I., 3–22, 33, 34, 39, 55, 58, 64, 113–132, 154, 191, 204, 211, 225–245
Orlikowski, W.J., 66
Ortega y Gasset, J., 32
Ostrom, E., 87, 105
Owen, R., 74

P
Papanek, V., 29, 30
Papert, S., 45
Parayil, G., 11, 12
Peacock, W.G., 183
Pearson, Y., 10, 12
Pérez-Foguet, A., 137, 138
Peter, F., 61

Petrescu, D., 219
Petterson, J., 183
Pinch, T.J., 107
Podesta, J., 103
Pogge T., 32
Prakash, A., 155

Q
Qizilbash, M., 6, 90
Qureshi, S., 7

R
Radder, H., 244
Rahim, S., 103
Ramachandran, D., 200
Ranciere, J., 218
Ratan, A.L., 11, 13, 20, 114, 129, 130
Reddy, A.K.N., 90–92, 94, 95, 99, 100
Rhodes, J., 129
Richardson, H.W., 183
Rickerson, W., 103
Riha, R., 211
Robeyns, I., 4–7, 16, 19, 34, 50, 58, 60, 61, 64, 114, 117, 122, 123, 126, 141, 144, 154, 177, 191, 193, 225
Roeser, S., 9
Rooksby, E., 14, 228
Rosales, J., 104
Rose, N., 40, 48
Rowe, W.D., 184
Roy, A., 219
Roy, R., 88
Rubbo, A., 9
Ruggero, C., 90, 103–105, 107

S
Sachs, W., 88, 98
Sahay, S., 12
Sanford, J.A., 226
Sanoff, H., 200, 206
Sant, G., 92
Saravanamuthu, K., 62
Saz, Á., 137, 138
Scattone, R., 104
Schacter, M., 163
Schneider, T., 204
Scholtes, F., 97
Schumacher, E.F., 115, 120, 137, 189
Scoones, I., 138, 139
Sen, A.K., 4, 5, 8, 11, 14, 27, 34, 41, 42, 49, 50, 54, 58, 60, 61, 64, 65, 78,

87, 88, 90, 95–102, 105, 122,
141–144, 146, 154, 155, 157,
163, 165, 168, 177–179, 191–193,
199, 225, 227, 228, 242
Serres, M., 52
Seward, C., 8, 124, 126, 244
Shadrach, B., 11
Shahani, L., 6, 7, 90, 191
Shannon, K., 205
Sherry, C., 104
Shiva, V., 138, 139
Simon, H.A., 194
Slovic, P., 184
Smith, M.L., 8, 124, 126, 244
Soja, E.W., 204
Sparrow, L., 77
Sparrow, R., 77
Stahl, B.C., 13, 15, 20, 21, 57–74, 124
Stairs, D., 189
Star, S.L., 28
Stemerding, D., 82
Stern, P.C., 105
Stewart, F., 6, 90, 101, 144
Stonich, S., 105
Strand, R., 74
Stubington, J., 197
Sumithra, G.D., 91, 92, 99
Sunstein, C., 184
Swart, R., 88
Swierstra, T., 82

T

Talyarkhan, S., 115, 116, 121
Tan, T., 68
Tao, J., 68
Tatano, H., 183
Taylor, C., 101, 106
Thomas, J.J., 11, 12
Thompson, M.P.A., 124
Tierney, J., 88
Tierney, K., 183
Till, J., 204, 211, 219
Toboso, M., 9, 16
Toly, N., 88
Turner, J.F.C., 213

V

Van Boeijen, A.G.C., 226
van den Hoven, J., 3, 9, 14, 27–36, 63, 168,
228, 229
van de Poel, I., 35, 174, 175, 177
van Gorp, A.C., 175, 177
Van Reijswoud, V., 115
Van Willigen, M., 183
Vaughan, D., 11, 12
Verbeek, P.P.C.C., 244
Vermaas, P.E., 231, 232, 242
Vine, E., 104
von Tunzelmann, N., 11
Vose, D., 176, 184

W

Wade, A., 205, 208, 215, 216, 218, 219
Walker, M., 142
Walsham, G., 12, 155, 164
Wang, Q., 11
Wang, Y., 103
Watts, M., 142
Weber, E.U., 105
West, C., 217
Wetmore, J.M., 176
White, L.J., 137
Whitley, E.A., 60
Wilhite, H., 88
Williams, R.H., 90, 91
Willoughby, K.W., 119–122
Wilson, K., 183
Winner, L., 31, 107
Wolff, J., 177, 178, 184
Wren, C., 163, 166
Wresch, W., 11

Y

Yu, J., 103
Yun, S.-J., 88

Z

Zheng, Y., 11–13, 15, 16, 20, 21, 57–74, 114,
115, 154–156, 164, 1247

Subject Index

A

Actor-network theory (ANT), 16, 124, 126, 127
Adaptive preferences, 5, 61, 165
Affective computing, 67, 68
Affiliation (capability of), 79, 80, 83
Afghanistan, 164
Ageing, 77
Agency
 collective, 17, 145, 146
 freedom, 60, 62, 64, 178
 individual, 59, 62, 102, 105, 106
 situated, 61–62
Agentive amplifier, 35
Agricultural extension, 116
Aid, 157–158
Ambient intelligence, 63, 67–70, 73
ANT. *See* Actor-network theory (ANT)
Architects, 174, 213
Architecture, 17, 92, 107, 205, 211, 215, 216
Australia, 196, 197, 199
Autonomy, 44, 65, 66, 70, 72

B

Bicycle, 5, 34, 49, 108, 227, 228, 234, 235, 238, 239, 242–243
Biopolitical/biopolitics, 40, 48
Biotechnologies, 10
Bodily integrity, 49, 81, 194
Bolivia, 249
Bush mechanics, 197

C

Capabilities
 approach, 5, 16, 18, 19, 34, 59, 64–66, 81, 90, 96–99, 101, 102, 117, 118, 141, 142, 146, 155, 182, 191, 193, 204, 208
 basic, 126, 193, 229
 central, 79, 193, 194
 collective, 144, 145, 147
 combined, 124, 229, 230
 external, 14
 human, 5, 7, 34, 115, 121, 122, 131, 229, 230, 242
 instrumental, 20
 list of, 79, 180, 192, 193
 meaning of, 83
 sensitive design, 58, 191, 229
 space, 205, 212, 215
Care
 e-, 83
 elderly, 78, 79, 81, 83, 84
 health, 10, 12, 20, 77, 199
 home, 80
 practice of, 83
Children, 40, 45, 46, 52, 114, 143, 197, 243
Choice, technology, 13, 119, 127, 155
Civil structures, 175
Collective
 agency, 17, 145, 146
 capabilities, 144, 145, 147
Collingridge dilemma, 67
Complex artefacts, 173, 175
Consequences, 49, 58, 63, 66, 176, 179, 183, 185, 190

Conversion
   factors, 16, 34, 50, 118, 121, 156, 161, 209, 228
Cornucopianism, 88, 100, 106
Cost-benefit analysis, 10
Critical theory, 17, 59, 60, 62, 66, 124
Culture, 52, 79, 114, 124, 197

D
DEFENDUS, 90–91, 94, 100
Democracy, 6, 92, 102, 206, 212
Democratic
   discourse, 17, 18, 63, 66
   technics, 89
Design
   capability sensitive, 58, 191, 229
   capability set, 197, 198
   capacity, 196
   code, 175, 176, 182, 185
   inclusive, 16, 226, 229
   participatory, 205, 206, 209, 211, 212, 215
   policy, 9
   practices, 226
   process of, 18, 140, 210, 217
   social, 226
   turn in ethics, 31
   universal, 226, 231
   urban, 211
Designers, 9, 29, 174, 234, 237, 244
Determinism, 128, 129, 210
Development
   projects, 137, 142, 146
   studies, 12, 13, 40, 81
   technological, 43, 64, 162
   urban, 206, 208
Digital
   extension, 116
   saturation, 46
Dignity, 78, 79, 81
Dissensus, 217, 218

E
Economic growth, 88, 98, 139, 148
Economics, 58, 78, 95, 177
Education, 46, 48, 97, 117, 195, 199
E-governance, 12
Elderly
   care, 78, 79, 81
   people, 79, 80
Electrification, 92, 145
Emancipation, 60, 61, 64, 65

Empathy, 68
Empowerment/empowering, 17, 58, 118, 122, 140, 141, 211
End-use, 90–93, 98, 99
Energy
   sustainable, 89, 90, 103
   utility, 89, 90, 103
Engineering
   civil, 18
   design, 130, 131, 174, 231
Engineers, 29, 34, 175, 181, 182, 184
Environmental conversion factors, 16, 160
Ethics
   applied, 31
   bio, 10
   of technology, 20, 78
Ethnography, 20

F
Fiction, 84, 85
Freedom
   of choice, 49
   instrumental, 193, 199
   intrinsic, 193, 199
   political, 101, 154, 157
   process, 212, 219
   product, 212, 214
Friendship, 82, 84
Functionings, 17, 19, 34, 49, 96, 97, 99, 100, 114, 130, 141, 155, 160, 164, 177
Future, 28, 42, 51, 67, 81, 82, 85, 91, 104, 143, 146

G
Germ-line engineering, 10
Ghana, 164
Good life, 32, 83, 84, 192

H
Health care, 10, 12, 20, 77, 199
Hermeneutics, 16
Housing, 184, 194, 196, 199, 215
Human
   agency, 17, 44, 45, 63, 65, 72
   capabilities, 12, 14, 22, 33, 34, 118, 121, 122, 126, 129, 148, 163, 226, 229, 242
   centred, 17
   development, 17, 58, 78, 96, 138, 146, 154
   diversity, 16, 17, 60, 63–65, 226, 232
   enhancement, 10
   rights, 14, 64, 141

Subject Index

Humanism, 44, 206
Hybrids, human and non-human, 43

I
ICT
  access to, 12
  ICT4D, 12, 20, 114
Ideology, 62, 63
Indigenous
  Australian, 197
  knowledge system, 197, 198
Individual, 19, 20, 49, 62, 64, 98, 102, 122, 126, 177, 181, 184, 191, 217
Individualism
  ethical, 126
  methodological, 126
  ontological, 126
Inequality, 42, 101
Information
  capabilities, 154–156, 160
  needs, 121
  systems, 58, 124
  systems research, 17, 19
Infrastructure, 28, 31, 104, 107, 155, 161, 175, 179, 196
Injustice, 10, 136, 207, 243
Innovation, 66, 104, 108, 139, 140, 190, 197
Institutions, 31, 40, 103, 144, 157, 159, 160, 192, 194, 208
Intelligent systems, 77, 79
Internet, 13, 28, 46, 51, 124, 128, 139, 155, 161

J
Justice
  distributive, 14
  social, 44, 88, 190, 220

K
Kurnell Desalination Plant, 196

L
Learning, 45, 48, 53, 147
Liberation
  from, 21, 40, 43, 47, 53
  within, 21, 40, 53, 55
Literacy, 116, 156, 160
Livelihoods, 12, 13, 116, 117, 130
Love, 77, 82, 84

M
Means-end, 81, 82
Mediation, 63
Megamachine organization (of society), 88, 103
Mobile phone, 8, 12, 196, 199
Models, 66, 68, 138, 175, 207
Moral imagination, 16, 82, 85
mp3, 115, 116, 118, 127

N
Neuroelectronics, 69, 70, 73
Non-human, 43, 44, 53, 83, 84
Normative judgments, 230, 232
Normativity of technical artefacts, 226, 232
N-tuple, 192, 195

O
Octavia Boulevard Project, 196
OLPC. *See* One Laptop per Child (OLPC)
One Laptop per Child (OLPC), 40, 45–47, 51, 197
Ontology, 8, 124, 212
Operationalization (of capability approach), 19
Opportunity(ies), 45, 49, 59, 60, 64, 65, 141, 144, 177, 184, 194, 199, 211, 212

P
Pakistan, 166, 167
Parliamentary strengthening projects, 153, 158
Participation, 6, 18, 62, 103, 116, 118, 142, 147, 155, 196, 205, 209, 212, 217
Performance, 158, 175, 176, 185
Personal conversion factors, 16, 34
Philosophy of technology,
  empirical turn in, 22
Podcasting, 119, 125
Political economy, 93, 103, 199
Political liberalism, 22
Politics of knowledge, 41, 44
Power, 13, 17, 41, 43, 60, 62, 63, 65, 102, 116, 124, 208
Practical action, 115, 118, 125, 137
Practical reason, 79, 194
Public deliberation, 6, 181
Public-private partnership, 164

Q
Quality of life, 78, 199

R

Rationality, instrumental, 232, 243
Research agenda, 20
Resources, 16, 34, 58, 62, 117, 121, 128, 140, 148, 156, 160, 161, 163, 166, 177, 178, 184, 228, 242
Responsibility, 141, 143, 156
Risk
  assessment, 10
  management, 176, 182
Robot, 67, 82, 84

S

Scenario, fictional, 82, 83
Science and technology studies (STS), 17, 122, 124
Settlement upgrading, 9
Social conversion factors, 34
Structure
  of living together, 102, 103, 105
  socio-historical, 62
STS. *See* Science and technology studies (STS)
Sufficiency, 89, 136
Sustainability, 102, 104, 105, 142, 175
Sympathy, 86

T

Technological development, 43, 64, 162
Technology(ies)
  acceptance model, 11
  appropriate, 86, 115, 119, 121, 131, 137, 190
  assessment, 66, 74
  choice, 13, 119, 131
  emerging, 67, 71, 115
  ends of, 59
  evaluation, 72
  information, 30, 78, 80, 82
  interpretive flexibility of, 60
  transfer, 137, 138, 148
Techno-moral
  change, 81
  imagination, 86
Telecasting systems, 160, 164
Telecentres, 11, 13
Television, 29, 164
Threshold (for capabilities), 6, 10, 181, 226, 243

U

United Nations Development Program (UNDP), 6, 138–140, 146, 148, 159
Urban
  design, 211
  development, 206, 208
  planning, 17
Use plan, 231, 232, 236, 238, 242
User capacities, 230, 231
Utility, 5, 28, 105, 154, 179, 184, 227

V

Values, 12, 30, 61, 62, 65, 91, 97, 99, 102, 105, 137, 143, 144, 147, 155, 161, 183, 199, 205, 209, 211, 212, 214
Value sensitive design, 9, 30, 63
Voting, 32, 156, 199

W

Well-being, 4, 7, 13, 60, 64, 78, 96, 117, 127, 131, 154, 177, 179, 185, 191, 195

Z

Zimbabwe, 115, 118, 127, 129

CPSIA information can be obtained at www.ICGtesting.com
Printed in the USA
LVOW012218271112

308951LV00013BB/548/P

9 789400 738782